Alexander Heinz
Folded Forms

NATURE AND ART

Nature and Art, they go their separate ways,
It seems; yet all at once they find each other.
Even I no longer am a foe to either;
Both equally attract me nowadays.

Some honest toil's required; then, phase by
phase,
When diligence and wit have worked together
To tie us fast to Art with their good tether,
Nature again may set our hearts ablaze.

All culture is like this; the unfettered mind,
The boundless spirit's mere imagination,
For pure perfection's heights will strive in vain.

To achieve great things, we must be self-
confined:
Mastery is revealed in limitation
And law alone can set us free again.

J.W. VON GOETHE (TRANSL. DAVID LUKE, LIBRIS 1999)

Alexander Heinz

FOLDED FORMS

Symmetrical and Playful Paper Designs

SCHIFFER
PUBLISHING

4880 Lower Valley Road • Atglen, PA 19310

Contents

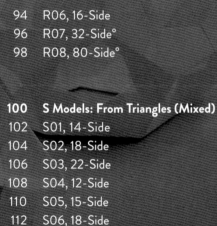

Preface

I first met Alexander Heinz at the annual convention of the Deutsche Gesellschaft für Geometrie und Grafik (DGfGG; German Society for Geometry and Graphic Design) in spring 2010. I had agreed to organize this convention at the former Imperial Abbey in Kornelimuenster, a historic part of Aachen where I had settled again following my retirement from the HFBK Hamburg. The event was to take place over several days, and I had instigated a competition in order to give it a special climax. All participants were invited to take part. The task was to contribute something from the subject area of geometry that—in the form of a physical or virtual model—would be able to call forth amazement, excitement, or a smile. The winner was promised prize money as well as the Phänomena sculpture, which was to be handed on as a prize cup of the DGfGG. Alexander Heinz was able win the competition by a large margin with his folded-paper polyhedra. This alone was reason and impetus enough for me to get to know his work and biography. The following years provided ample opportunity to do so.

On the road to a later convention of the DGfGG in Dresden, driving together in my car, we realized we had a lot in common. We both had a background in a practical craft, but the material with which we had been trained to work was different. For Alexander Heinz as a bookbinder, it was paper; for me as a carpenter, it was initially wood, later also metal. During our conversation, we pondered whether craft activities had any influence on people's abilities to grasp geometrical concepts. We agreed that any creative craft activity such as measuring, forming, folding, bending, joining, etc. always contains an element of geometry. During the actual work process, this connection is usually hardly noticed. It instead happens in the subconscious, but it means that geometrical insights are acquired and retained. Over the course of my teaching career at the HBFK Hamburg, I found that students with a background in traditional crafts generally had significantly fewer problems in understanding descriptive geometry. They had already developed their spatial imagination, which is indispensable for its application of descriptive geometry. But there's another aspect of a background in a traditional craft—and that is the intimate familiarity with a given material. This is an advantage especially for those who want to work creatively with a particular material. The more one knows its characteristics and possibilities, the greater are the possibilities to discover and dare something new.

Another thing we have in common is the fascination we both have with regular solids. Since my time at university, I have repeatedly built regular and semiregular polyhedra. In the beginning, I used cardboard, drawing and cutting out flat patterns and gluing them together along flaps. Later, I used the ball-and-stick technique or aluminum, cut on a bevel and joined with epoxy resin. Eventually I experimented with cut planes, which were put vertical to the rotational axes and cut from stainless steel with a laser cutter. But polyhedra, built from modular paper elements and simply pushed together, were a completely new territory for me. Only someone intimately familiar with paper as a material can have such a genius idea.

When we show polyhedra constructed in different ways, their appearance changes, and different characteristics become evident. If the surface is closed, the solid appears monolithic. If only corners and vertices are shown (as in the ball-and-stick technique), their connections dominate. If the solid is constructed from cut planes joined vertical to the rotational axes, their symmetries come to the fore. Alexander Heinz's folded and joined polyhedra emphasize corners and vertices, while the faces generally remain without material, appearing as funnel-shaped recesses so that one can see right through the entire solid.

In his first book, *Folding Polyhedra*, Alexander Heinz systematically presented the construction of all Platonic, Archimedean, and Catalan solids with instructions for his readers to replicate the models. In this follow-up volume, he retains the underlying principle of construction using "horse" and "rider," but the stringent geometry of polyhedra is relinquished. This means that your imagination is invited to run wild without having to follow the strict rules of symmetry. In this way, numerous new forms can arise as a development from regular polyhedra.

May this book be a success, and its readers experience a maximum in creativity!

Friedhelm Kürpig
Prof. (retired) for Constructive Geometry
HFBK Hamburg
(1978-2007)

An Introduction

A Folding Technique: Joining Horse-and-Rider Modules

In the beginning, there is a small stack of twelve identical square pieces of paper on the table in front of you: each square is folded twice parallel to the edges, turned over, and then folded twice diagonally. When you then circle the center of the piece of paper with your finger, you will trace mountain and valley folds alternatingly, as shown in the illustration above. Mountain folds are indicated by a continuous red line, valley folds with a dotted gray line. Each project in this book is accompanied by illustrations. Two pieces of paper joined together using the horse-and-rider principle form one module: the upper piece "rides" the horse below, as shown in the illustration above (*right*).

A total of six modules can be joined together step by step to form a spatial body or solid, called octahedron (see illustrations, page 10). The name "octahedron" means "having eight faces," counting the eight triangular "windows" between modules. The horse-and-rider principle is simple and genius at the same time—and serves as a basis for all the forms shown in this book. Further experiments suggest themselves, following a first successful result such as changing the starting conditions. Using squares results in different spatial forms (for example, models of the N, O, and P series). Individual valley folds can be omitted to this end.

On page 19, you find an overview of all modules used in this book. Each module has been represented in a particular color according to the types of valley folds used. This color has been listed in the table on page 19 and is explained in detail in the instructions starting on page 22.

The modular approach can also be tried out with triangular pieces of paper, leading to the models of the Q, R, and S series. Here, too, certain valley folds can be omitted or only partially executed. Additionally, the combination of triangular and square modules leads to satisfying results (cf. models of the T and U series). Even pentagonal and hexagonal modules can be realized. Finally, further solutions are possible that are not considered in this book.

The size of the individual modules has to correspond so that the modules can be joined correctly. Detailed measurements can be found at https://www.haupt.ch/faltformen.

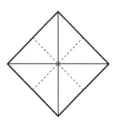

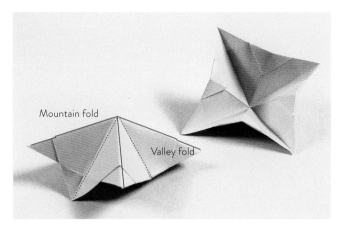

Mountain fold

Valley fold

Mountain fold

Valley fold

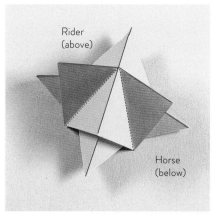

Rider
(above)

Horse
(below)

Folding the tips of the rider around the vertex
of the horse

See the following link for an instructional video:
https://www.haupt.ch/faltformen.

A Modular Series of Forms: Regular and Semiregular Polyhedra

The first convincing results spark off an interest in the spatial laws behind them: which types of spatial forms can be created using these modules. Two paths suggest themselves in order to answer this question. On the one hand, the forms arise in a playful way through trial and error. This is the approach followed in *Folded Forms*. On the other hand, it is possible with this method to work toward a particular preconceived form.

If you know the treasury of forms of the five regular and 30 semiregular spatial forms (Platonic, Archimedean, and Catalan solids)*, it may be tempting to find out which of these forms can be realized using the horse-and-rider technique. Some solutions for complex spatial forms are surprisingly simple. However, some seemingly simple spatial forms, such as the cube, can pose some initial difficulties. If you persist and follow your line of inquiry, you'll gain insights into the geometry and its lawfulness that will lead you to convincing solutions. In this way, you'll be able to increase your repertoire of forms. The first big goal then is to realize all the regular and semiregular spatial forms using the horse-and-rider technique. This already determines the appearance of a desired form where its vertices are concerned. The technique that will yield the finished model is, at least in its principles, also known. The solution that lies in between these two points, however, initially remains shrouded in darkness. Finding this solution is a bit like navigating city traffic without a map: "I know where I start, and I know where I want to get to. But I have to find the way while I'm going." Eventually, all regular and semiregular forms can be realized. The book *Folding Polyhedra* treats exactly that process in its chapters A to I**. The later chapters in *Folding Polyhedra* lead to the approach that has become the focus of this book—folding polyhedra in the space between strictly symmetrical spatial structures and a free, playful experimentation and combination of the different folded modules, which are shown on page 19.

* See also glossary, page 166.

** As a general note: *Folding Polyhedra* uses letters A to M for its model series, which are continued with the series N to U in this book.

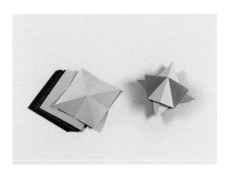

First steps: Folding the paper bases and creating a module from horse and rider

Joining the modules into a threefold ring joint

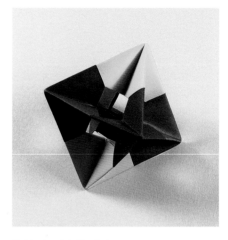

Finished model: Octahedron

Further Forms

Free experimentation results in the insight that it is possible to create a large number of less symmetrical spatial forms with the horse-and-rider technique by combining triangles, squares, pentagons, and hexagons. The final form, however, emerges only when the last module has been fitted tightly into the model. Initial joints must fit; others are tried out. If things don't fit, alternatives have to be found. Many experiments lead to a dead end. Some beautiful experiments that cannot be completed, such as the model T07 (24-Side), have still been included in this book as torso* (see page 137). All models can be taken as the results of coincidental finds that nevertheless have a geometrical explanation. Usually, there is a relationship between the regular and semiregular polyhedra. Insofar as geometry is a spiritual affair, we find relationships in spirit between the models. This approach that took its beginning in *Folding Polyhedra* (chapters J to M) is now continued in *Folded Forms,* using the subsequent chapter designations N to U**. It should be noted, however, that this second volume can stand on its own: no model is repeated, and no prior experience or knowledge is presupposed.

Free Folded Forms

Thus, the book presents different folded spatial forms and their construction. These arise from the possibility to freely combine triangles, squares, pentagons, and hexagons, which are joined to form modules using the horse-and-rider technique. The models have been grouped according to similarities in their construction. For models N, O, and P, you predominantly need square modules. For models Q, R, and S, mostly triangular models are used. The models of the T and U series combine mainly triangular and square modules (as well as pentagonal and hexagonal modules).

The color code helps to see which type of fold is needed where in a model. The overview (table) on page 19 shows this clearly: a blue square has all possible red mountain folds as well as all possible gray-dotted valley folds. The yellow square also has all possible red mountain folds, but only one gray-dotted valley fold. The orange, the light-green, and the dark-red squares show yet a different pattern of folds. In a similar way, this is true for all other modules: each color marks a different selection of necessary valley folds.

It's characteristic of the approach presented here that aspects of different contexts that are usually regarded as separate are combined. This is, for example, expressed in the didactic hints at the end of the book (see page 162).

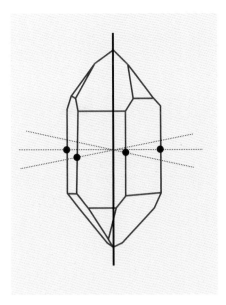

Clear quartz crystal with threefold main symmetry axis and three two-fold subsidiary symmetry axes

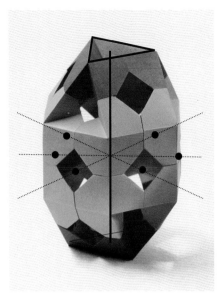

Model O04, 20 Side, with threefold main symmetry axis and three two-fold subsidiary symmetry axes

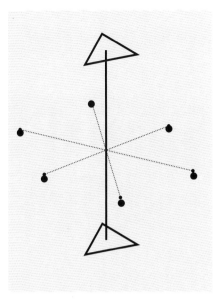

Illustration of threefold main symmetry axis and three two-fold subsidiary symmetry axes

Looking at symmetry axes* (see page 12) provides an additional point of orientation when attempting to describe, understand, and differentiate individual models. In crystallography, symmetries play an especially important part in the differentiation, and classification of different spatial forms has taken over this principle (see illustration above). A clear quartz crystal, for example, has a three-fold main symmetry axis. During a complete rotation along this axis, the same picture of its facets appears subsequently three times when seen from one point of view. A clear quartz crystal additionally has three two-fold subsidiary axes. During a complete rotation along these axes, the same picture appears only twice.

Understanding Symmetries

The same is true for the model 20-Side (see page 40), illustrated above. If you find it difficult to understand the matter immediately, you can simply start by building the model. While turning the model this way and that in your hand, the idea will soon become clear. The construction of the models is always the main thing; the symmetries merely serve their differentiation and should be regarded as an additionally offered dimension to understanding them. The illustration at above right shows an abstract view of the symmetries. Each set of instructions has a corresponding illustration of this type. This will be explained in detail on the following pages. Finally, it's obvious that the Western tradition of the geometry of

polyhedra forms a union with the Eastern materiality of paper and the folding techniques of origami. Two different worlds unite in this modular approach to the construction of these models. Similar aspects are reflected in other perspectives: as the introductory sonnet by Goethe puts it it so well, the construction of these models combines craftsmanship, a striving for knowledge of the (natural) sciences, and the creative aspect of art and nature into a whole.

And so we see that if you follow something of your own choice to the very bottom of its matter, with time many relationships that govern the world—seemingly paradoxically—reveal themselves.

* See also glossary, page 166.

** Most of the models in this book are technically less demanding than many of the forms in *Folding Polyhedra*. It may be of interest to look into *Folding Polyhedra* since it provides a substantial introduction to the geometry of polyhedra, as well as an introduction to the cultural and historical background of the subject.

Spatial Structures and Symmetries

Spatial Structures

Spatial structures play an important part in many general aspects of daily life as well as in technology and science: in crafts, immediately when it comes to the material; in mathematics, in terms of numbers and theory; and in geometry, principles of construction are a focus. Generally, spatial structures always exhibit the following four geometrical elements: corners (points), vertices (straight or curved lines), flat or curved planes (faces), and spaces (volumes). While scientists use spatial structures for purposes of exact orientation, craftspeople apply and build these structures in practice. Artists treat these structures more playfully and use them as media of expression. This book combines all of these approaches.

Point/Corners, Straight/Vertices, Plane/Faces, Space/Volume

Any spatial form can be described according to the number of its corners, vertices, and faces. Any type of cube, for example, is a form with six faces, eight corners, and twelve vertices. Its realization can differ widely depending on the materials used: molded from concrete, built from poles as playground equipment, flat surfaces joined into a box, or merely imagined as an immaterial thought model in the form of a cloud of points with eight points (with exactly defined distances between the points). Depending on which aspects are of importance for the intended construction—whether the intended form is concrete and material or a thought model—individual characteristics can be emphasized. These are decisive when it comes to looking at and naming the outcome.

A carpenter working with solid wood or a builder casting a cube from concrete will predominantly focus on the volume of a spatial form. But if boxes or simple living spaces are being constructed, the focus lies on the planes. A piece of playground equipment built from poles emphasizes the vertices. In chemistry, the scientist will consider the corners or knots of a spatial structure in his spatial thought models. All these perspectives can be used to describe folded forms. This book, however, focuses on the type and number of modules, as well as the number of ring joints (or the "windows" between the modules; see illustrations on pages 10 and 11). The characteristic symmetries—as used in crystallography to describe different spatial structures—will help distinguish the folded forms.

Illustrations for models with low symmetry with only one main symmetry axis (no subsidiary symmetry axes)**

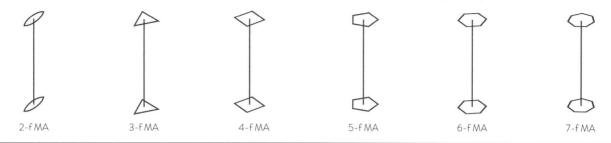

| 2-f MA | 3-f MA | 4-f MA | 5-f MA | 6-f MA | 7-f MA |

Illustrations for models with higher symmetry with one main symmetry axis and several subsidiary symmetry axes

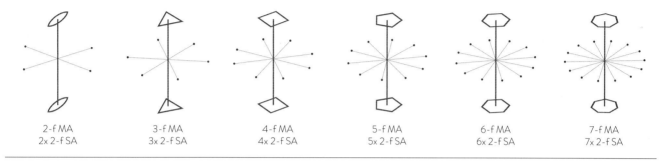

| 2-f MA | 3-f MA | 4-f MA | 5-f MA | 6-f MA | 7-f MA |
| 2x 2-f SA | 3x 2-f SA | 4x 2-f SA | 5x 2-f SA | 6x 2-f SA | 7x 2-f SA |

Crystallographical Structures: Rotational Symmetry Axes

Crystalline forms are systematically described according to their symmetry axes. Any regularly formed crystal can be turned around one or more axes. A 360° turn around can be imagined as seen in the starting position (a simplified version is shown on page 11). The vertical axis is threefold and is called the main symmetry axis. Additionally, there are three radial subsidiary symmetry axes, which are only twofold (see page 11). We can see that crystals, as the most stable, unchangeable forms of nature, are classified using a method that, when practically applied, asks for a high degree of inner mobility. We use these symmetry axes as a possible way to differentiate the folded forms. One of the essential foundations for this approach was given by the Danish scientist (later also priest and bishop) Nicolaus Steno, who discovered in 1669 that the angles between crystallographically equal planes were always the same size. In 1801, French mineralogist René-Just Haüy (1743–1822) introduced symmetry as the term on which formal definition in crystallography rests, as we have shown in a simplified way for the clear quartz crystal and applied to a folded model (see page 11). We can't here go into the very detailed aspects of crystallography (further reading on page 169).

Simple Symmetries

Most models in this book correspond to the symmetry principle of the clear quartz crystal, with one main symmetry axis and the respective number of radial subsidiary axes. Related forms can also exhibit 4-, 5-, and 6-fold rotational symmetries. In this way, several very similar models can be built that differ chiefly in the number of their rotational symmetries. The number of rotational symmetries is given in the instructions for each model (see illustration above). Rotational symmetries always run axially: these axes run through diametrically opposed corners of a spatial form or through the center of two facing vertices or through the center of two facing planes. In simple symmetries we distinguish between symmetries with only one multifold main axis (*upper row*) and symmetries with additional twofold subsidiary axes that run vertical to the main axis (*bottom row*; subsidiary axis is shown horizontally).

* See also glossary, page 166.
** MA = main symmetry axis
   SA = subsidiary symmetry axis
   f = fold

Symmetries in a tetrahedron | Symmetries in a cube/octahedron

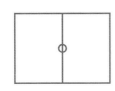

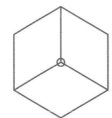

3 x 2-f A

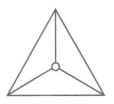

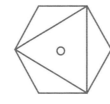

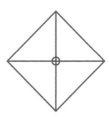

4 x 3-f A   6 x 2-f A   4 x 3-f A   3 x 4-f A

Complex Symmetries in Polyhedra

Highly symmetrical forms are a separate category. Here, two or three axes with different symmetries interlock. These are simply called symmetry axes in the following. Only three out of the models in this book show these complex symmetries. If seen from this point of view, these three models are the apex of what is possible in terms of symmetries in spatial forms.

Trying to illustrate all axes according to the illustrations on page 13 in complex symmetries in polyhedra soon becomes confusing due to the large number of axes. In this context, we show only which symmetries are present for one corner, one face, or one vertex (see illustration above). These then have to be applied to all corresponding corners, faces, or vertices as appropriate.

Symmetries in a Tetrahedron

The tetrahedron is a form consisting of four equal triangles. If we look at its vertices, we find twofold symmetry axes. The tetrahedron has six vertices—since the symmetry axes always run through two vertices that face each other, the tetrahedron has three 2-fold symmetry axes. Looking at one of the faces as well as looking at one of the corners, the tetrahedron reveals a 2-fold symmetry. Each corner of a tetrahedron is opposite one of its faces. The tetrahedron has a total of four 3-fold symmetry axes. Examples for tetrahedron symmetries can be found at above left.

The model R06, 16-Side, has the same symmetries as the tetrahedron (see page 94).

Symmetries of a Cube and Octahedron

A cube is a form made up of six equal squares. Looking at its faces, we find 4-fold symmetry axes. The cube has three of these. Three vertices and three faces meet at each of its corners—here we see 3-fold symmetry axes, of which the cube has four. Looking at its vertices, a total of six 2-fold symmetry axes are revealed.

Similar characteristics are exhibited by the octahedron (see page 10). Looking at its triangular faces, there are 3-fold symmetry axes. Of these, the octahedron has four. At each of its corners, four vertices and four faces meet—here we see 4-fold symmetry axes, of which the octahedron has three. Looking at its corners, there are a total of six 2-fold symmetry axes. You will quickly grasp that cube and octahedron look different, yet in terms of the number and foldness, they have the same symmetries. Examples for the symmetries of the cube and octahedron can be found above, right next to those of the tetrahedron.

Symmetries of dodecahedron and icosahedron

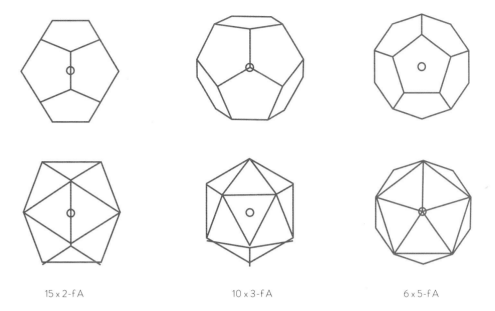

| 15 x 2-f A | 10 x 3-f A | 6 x 5-f A |

The model R07, 32-Side, has the symmetries of both cube and octahedron (see page 96).

Symmetries of a Dodecahedron and Icosahedron

The dodecahedron is a form consisting of twelve equal pentagons (see page 146, figure 5, model at the far right). Looking at its faces, we see 5-fold symmetry axes. The dodecahedron has six of these. Three faces and three vertices meet at each of its corners—here we find 3-fold symmetry axes, of which the dodecahedron has ten. Looking at its 30 vertices, we see a total of fifteen 2-fold symmetry axes.

Similar characteristics are shown by the icosahedron (see page 146, figure 5, model on the left). Looking at its 20 triangular faces, we see 3-fold symmetry axes. The icosahedron has ten of these. At each of its corners, five vertices and five faces meet— here we see 5-fold symmetry axes, of which the icosahedron has six. Looking at its 30 vertices, we see a total of 15 2-fold symmetry axes. In this case, too, you see that the dodecahedron and the icosahedron are different in appearance, but the number and foldness of their symmetries are the same.

The model R08, 80-Side, has the symmetries of dodecahedron and icosahedron (see page 98).

There are many examples for crystal forms with simple symmetries with one main axis, as well as some with additional 2-fold subsidiary axes. Apart from an entire range of symmetry systems that have not been included in this book, there are crystal forms exhibiting the symmetries of the cube and of the octahedron. It should be noted that not every crystal has to be formed as perfectly in its symmetries as is possible according to the laws of the mineral kingdom. The highly complex symmetries of the dodecahedron and icosahedron are not found in the mineral kingdom, but in the realm of viruses.

* *Folding Polyhedra* includes several of such forms.
** A = axis; f = n-foldness

Practical Considerations: Cutting, Folding, and Colors

Paper Quality Required

Well-sized, fully colored, and form-stable paper with a weight of 120g/m² is best suited for the projects in this book—for example, f.color® smooth, Efalin smooth, or Surbalin smooth. Such paper is often used for covering the boards in books or as flyleaf. All models in this book have been constructed using f. color®. A firm, well-sized, form-stable, universal-use paper with a weight of 120g/m² is a good alternative. It has to feel stiffer than ordinary printer paper, more like paper commonly used for high-quality maps.

Cutting Squares and Triangles from Strips of Paper

When cutting the required paper shapes, take several (five to ten) layers of paper at once and staple them together firmly. In this way, you can cut several basic areas at the same time.

To cut squares and triangles, first cut individual strips out of the sheets of paper. The width of the strips must correspond to the desired length of the edges. For triangles, the strips must be as wide as the desired height of the triangles. If you know only the length of the edges of the triangles, you will have to make some calculations to construct the correct height.

The strips for squares are cut at right angles to the final measurements. With a guillotine, this task is easy to accomplish,

but it can also be done with a set square, a cutter, and a cutting mat. For triangles, the strip must be cut at a 60° angle. Don't forget to turn over the strip after each cutting made; otherwise you end up with diamonds (fig. 2).

Cutting Pentagons and Hexagons Using Templates

For the pentagons and hexagons, it's best to make copies of the templates (more on page 19) and use them as stencils. To do so, put the desired template on the paper. Using a bookbinder's awl, punch a hole at each corner. Then use a cutter to cut a line from one hole to the next. It's possible to cut several layers of paper in one go, but be careful to hold the cutter absolutely vertical at all times. Otherwise, the bottom layers will have slightly deviating measurements.

Size and Color of the Folding Papers

All required basic paper shapes are listed with their colors and characteristics in the table on page 19. To begin with, it's advisable to follow these specifications. Later, you may wish to experiment with other specifications, depending on your intention or following certain practical considerations.

If you would like to construct smaller models (these are easier to keep and transport), you can use a Xerox machine to make the templates smaller. If you want to construct models that are a

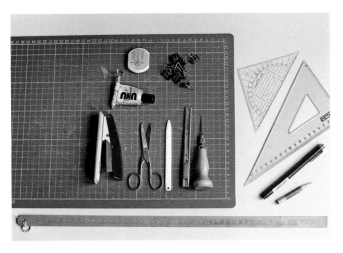

Fig. 1: Required tools

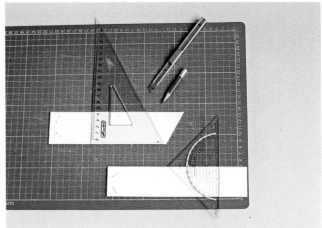

Fig. 2: Cutting triangles and squares

combination of several different folded forms, you have to decrease the templates proportionately (i.e., by the same factor) to the original size. The same applies when making larger models.

Using Glue
Most joints are stable simply because they interlock; the modules keep each other in place. In some cases, it makes sense to secure the folded-over tips of the riders on the bottom of the horses with a dot of glue (especially with pentagonal and hexagonal modules) or by gluing the joints together (or both). If this is the case, it has been included in the respective instructions below. Clear, solvent-based glue has proven reliable, but some people prefer to use PVA glue instead.

Supports
Where pentagonal and hexagonal modules are joined together, the joints can become unstable. In addition to glue, it's helpful to use supports—narrow strips of folding paper, folded at the center, or even projector foil cut to the width of the joint, make optimal supports that lend excellent stability to any joint. Respective templates can be found at https://www.haupt.ch/faltformen.

Size
The previously cut triangles, squares, pentagons, and hexagons are used to create triangular, square, pentagonal, or hexagonal modules (for more information, see table on page 19). To combine any module with another freely, the joining parts must have the same width. Therefore, you need to cut the shapes according to the measurements given. If you choose a different size, remember to increase or decrease the size of all your pieces of paper in the same ratio compared with those given in the table. You can find all templates here: https://www.haupt.ch/faltformen.

Necessary Mountain and Valley Folds in Different Models
Without exception, all the models presented here need all possible mountain folds, as can be seen in the table on page 19. The angles between two mountain folds determine the angle at which two polyhedron vertices meet in the finished model. This angle between two mountain folds cannot be increased. It can be decreased only by adding a valley fold between the mountain folds. This makes the angle between two mountain folds variable within certain limits. They can thus be individually used as needed. The necessary folds can be executed by hand or by using a folding bone. Take your time when folding each horse and rider, as well as when joining the modules together. Take care to execute the folds accurately and exactly.

Fig. 3: Transferring the measurements from template to folding paper

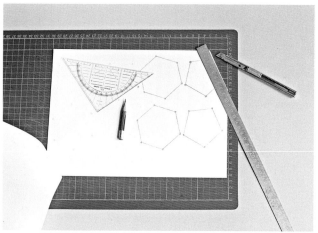

Fig. 4: Cutting out pentagons and hexagons

Special Valley Folds and Color of the Modules

Depending on the type of module, only those valley folds listed in the table are needed. It's helpful to color-code folded modules according to the number of their valley folds. Triangles with all possible valley folds are shown in light blue (see facing page). If only two or three out of all possible valley folds in a triangle are folded, we are using turquoise-colored paper. Pink paper is used for triangles with only one valley fold. The same approach is applied to square papers. Here, a total of five versions are possible, all of which are used in the models presented in this book. In the pentagonal and hexagonal folded modules, all five possible valley folds have been used in all the modules. The forms created with our approach emerge most clearly when using just white paper, as the double-page spreads preceding each chapter illustrate. However, orientation is much easier when modules are used whose color indicates the number of necessary valley folds.

Deviating Measurements for Horse and Rider

For a very small number of models, it's necessary to deviate for certain modules from the standard measurements used elsewhere: horse or rider (or both) then have a different size from that given elsewhere. These have been clearly marked with the special character ° (for example, R07, 32-Side, on page 96 and R08, 80-Side, on page 98). You can find the exact measurements for the modules in question here: https://www.haupt.ch/faltformen.

Angular Ring Joints* and Model Terminology

All modules interlock in closed, angular rings whose sides or corners can be counted. A five-fold ring joint in a model forms an imagined pentagonal face. If you add up all the ring joints in a model, you get the name of the model (e.g., 19-Side; see pages 20–21). Models that look completely different at first sight still have the same number of faces but have completely different symmetries.

Difficulty and Construction Time

Constructing your own models depends to a high degree on individual talents and experience and sometimes on daily performance. All models in *Folded Forms* have an indication of their difficulty and the time needed for the construction. These indications are to give you an orientation. It is generally advisable to start with simple models and proceed incrementally and with growing success to the more difficult forms.

*See also glossary, page 166.

Form	Page	Color	Valley Folds	Illustration Rider	Picture Rider	Illustration Horse	Picture Horse	Horse/Rider	Picture Module
triangle	79		all valley folds						
	73		1 of 3 valley folds						
	85		2 of 3 valley folds						
square	25		all valley folds						
	55		3 of 4 valley folds						
	35		2 of 4 valley folds (every second one)						
	25		2 of 4 valley folds (two adjacent ones)						
	35		1 of 4 valley folds						
pentagon	27		all valley folds						
hexagon	29		all valley folds						

Layout of the Model Pages

Note

This page is an example of an overview of the layout for the constructions of the models in the in the instruction part that follows. Each double-page spread follows the same pattern. This facilitates orientation when constructing your own models.

Hint regarding figs. 5–7

Measurements to scale of the basic pieces of paper for horse and rider can be downloaded at https://www.haupt.ch/faltformen as a 1:1 printer/copy template. If a model requires only a single type of module with horse and rider of the same size, it can be constructed in any size desired.

1

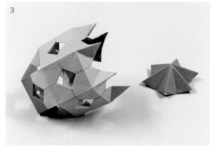

10

1 **Model Number**
Letter: Designates a particular group of models (based on technical characteristics; e.g., models based on triangles, squares, pentagons, and hexagons)
Number: running model number

2 **Model Name / Technical Characteristic**
Example: O05, 19-Side. Belongs to the O models, mostly based on orange and yellow squares.

3 **Model Pattern / Main Axis**
Example: 19-Side: This model has a three-fold main axis, here shown vertically with triangular ends.

6 Basic Paper Shape for Horse (Schematic Illustration)

Horse papers are covered by rider papers and for the underside of a module.

7 Pictogram

Rider always the upper part of the module

Horse always the lower part of the module

8 Number of parts

The number given indicates how many sheets of horse and rider papers, respectively, are needed for the construction of this model. See also the overview on page 19.

9 Step-by-Step Instructions

To be followed according to the way explained for the example of the 19-Side. First steps facilitate the beginning; further steps follow from these. If applicable, similar models are referred to. Finally, an approximate estimation of the time needed.

O05   19-Side   O-Models: From Squares (Yellow, Orange)

19-Side ◼2

First Steps

You will need 6 orange squares (3 x horse and 3 x rider) as well as 12 yellow squares (6 x horse and 6 x rider). Additionally, you will need a further 12 light green squares (6 x horse and 6 x rider) as well as 2 moss green hexagons (1 x horse and 1 x rider). Execute mountain and valley folds according to the illustrations on the right. Join horse and rider to form 3 orange, 6 yellow, 6 light green and one moss green module.

Joining the Model

Join 3 orange modules to form a triangular ring joint (fig. 1). Continue from the orange modules with 6 yellow modules (fig. 2). Then join 6 light green modules to these, in a similar way as in model N03 (fig. 3). The six light green tips are topped with a concluding moss green module (fig. 4). For this model, you can access video instructions using the link: https://www.haupt.ch/faltformen.

Similar Models

As can be seen in fig. 5, O05 is a combination of N03 and O04.

Difficulty: Moderate

Cutting: some challenges, moderately challenging construction; time required: approximately 1 hour.

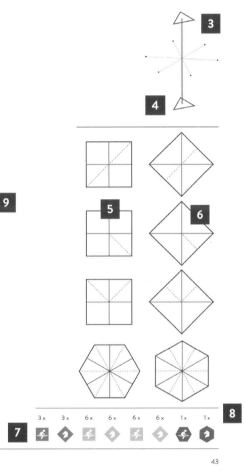

3x   3x   6x   6x   6x   6x   1x   1x

43

4 Template Module / Subsidiary Axes

This model has three 2-fold subsidiary axes, here shown radially around the vertical main axis.

5 Basic Paper Shape for Rider (Schematic Illustration)

Rider papers always lie on horse papers and are visible as the upper part of the module.

10 Step-by-Step Assembly Instructions

Assemble the model in the sequence indicated in the pictures. These steps follow the step-by-step instructions.

N Models:
From Squares
(Light Green)

Each of the four N models has square modules with two valley folds (light green). Two modules together always form a twin module that can be further joined to form tubular units of 2 x 4, 2 x 5, and 2 x 6 modules. In this way a 4-fold (N01), a 5-fold (N02), and a 6-fold (N03) main symmetry axis results. The 2-fold subsidiary symmetry axes are at a right angle to the main axis: four in model N01, five in N02, and six in N03. In the N models, the number of ways (or radials) of the main axis is the same as the number of 2-fold subsidiary axes.

There are limits to the simple continuation of counting up or down, both in practical and in geometrical terms. The models of the 9-Side (3-fold main axis), the 21-Side (7-fold main axis), and the 24-Side (8-fold main axis) when constructed according to the same principles can be realized only as a torso (truncated).

With the higher model numbers, there are increasing cross-references (construction, modules, symmetries, topology) among the models. These are occasionally highlighted in the instruction text, especially where similarities are particularly obvious or notable. Each model in this chapter is a self-contained construction series. Some arrangements of modules reappear with models in later chapters.

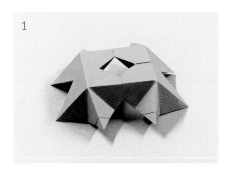

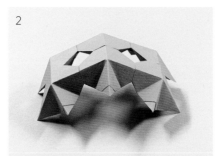

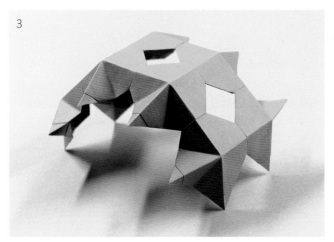

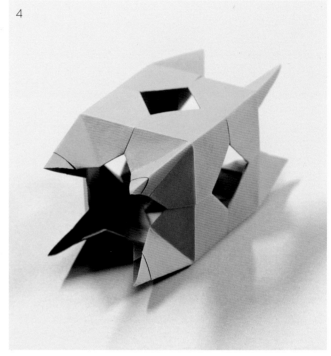

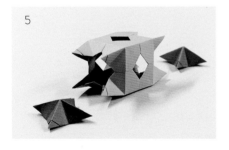

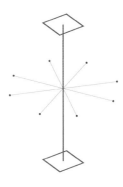

12-Side

First Steps

You will need 16 light-green squares (8 x horse and 8 x rider), as well as 4 dark-blue squares (2 x horse and 2 x rider). Execute mountain and valley folds according to the illustrations on the right. Join horse and rider pieces to form 8 light-green and 2 dark-blue modules.

Joining the Model

Join four light-green modules to form a first square ring joint (fig. 1). Join two further light-green modules to this (fig. 2) and, in the same way, add the last 2 light-green modules. A tubular shape results, with 4 tips at each of the two openings (figs. 3 and 4). One dark-blue module is then attached to these light-green tips on each side. The resulting 12-Side (fig. 5) has a 4-fold main symmetry axis.

Similar Model

Using 6 light-green modules, you can construct a similar form (9-Side) with a 3-fold main axis. This, however, can be realized only as a torso or truncated version (fig. 6, shown alongside N01).

Difficulty: Easy

Simple cutting, simple construction; time required: approximately 45 minutes.

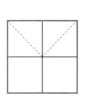
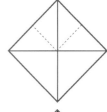

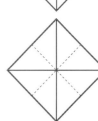

8 x   8 x   2 x   2 x

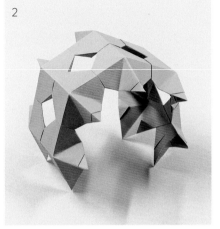

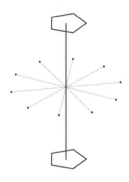

15-Side

First Steps

You will need 20 light-green squares (10 x horse and 10 x rider) as well as 4 violet pentagons (2 x horse and 2 x rider). Execute mountain and valley folds according to the illustrations on the right. Join horse and rider to form 10 light-green and 2 violet modules.

Joining the Model

Join 6 light-green modules to form the first square ring joints (fig. 1), then add the remaining light-green modules (fig. 2) to create a tubular shape with 5 tips at each opening (fig. 3). Join a violet module at each end to these tips. The resulting 15-Side (fig. 4) has a 5-fold main symmetry axis. Fig. 5 shows N01 and N02 together.

Difficulty: Easy

Simple cutting, simple construction; time required: approximately one hour.

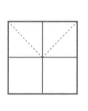
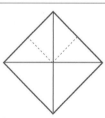

10 x   10 x   2 x   2 x

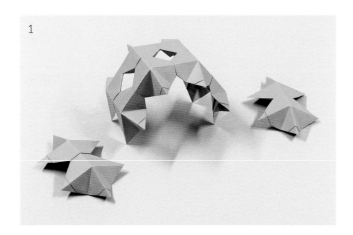

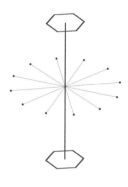

18-Side

First Steps
You will need 24 light-green squares (12 x horse and 12 x rider), as well as 4 moss-green hexagons (2 x horse and 2 x rider). Execute mountain and valley folds according to the illustrations on the right. Continue to create 12 light-green and 2 moss-green modules with the horse-and-rider squares.

Joining the Model
Take 8 light-green modules and join them to form first square ring joints (fig. 1), then add the remaining light-green modules to create a tubular shape with 6 tips at either end (fig. 2). Join a moss-green module to these light-green tips. The resulting 18-Side (fig. 3) has a 6-fold main symmetry axis. Fig. 4 shows N01, N02, and N03 together.

Difficulty: Easy to Moderate
Cutting: some challenges; simple construction; time required: approximately 45 minutes.

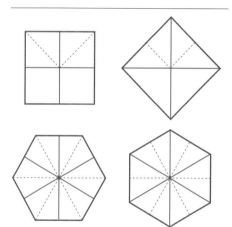

12 x   12 x   2 x   2 x

21-Side
(Torso/Truncated)

First Steps
You will need 28 light-green squares (14 x horse and 14 x rider). Execute mountain and valley folds according to the illustration on the right. Join horse and rider to create 14 light-green modules.

Joining the Model
Join 8 light-green modules to form the first square ring joints; add the remaining light-green modules (fig. 1), creating a tubular shape with 7 tips at each opening (fig. 2). The resulting form remains truncated, or a torso (21-Side), without heptagonal modules to complete the shape, with a 7-fold main symmetry axis.

Similarities with Other Models
Using 16 light-green modules, you can create a similar shape (24-Side) with an 8-fold main symmetry axis, which also can be realized only as a torso (fig. 3; N03 is at the far right of the picture).

Difficulty: Easy
Simple cutting, simple construction; time required: approximately 45 minutes.

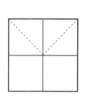 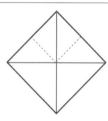

14 x   14 x

O Models:
From Squares
(Yellow, Orange)

All 10 O models have square modules, with one valley fold (yellow) or two valley folds (orange). With these models, as with the N models before, you will come across tubular models with two completing end parts at the sides. O01 to O03 form a separate series that could also include the cuboctahedron*. The common basis of O04 and O05 is a hexagonal tube forming a central part with different extensions at the ends (see also the R models, page 83). The 3-fold orange modules that complete the models allow for only one 3-fold main symmetry axis.

O06 to O08 have only main symmetry axes and no subsidiary symmetry axes. O09 can be built in two directly symmetrical (mirror) forms (chiral)** and is similar in the kind and number of modules to the rhombic cuboctahedron top***. Owing to the low symmetries, these models look completely different when viewed from a different perspective. This makes them particularly attractive—their full form emerges only when looked at from all sides. It is almost impossible to capture this in photography.

In most cases, it helps to stabilize the folded tips of the rider parts (underneath the horse parts) with a drop of glue, and likewise the joints between individual modules.

* A02 in *Folding Polyhedra* / ** See also glossary, page 166 / *** J03 in *Folding Polyhedra*

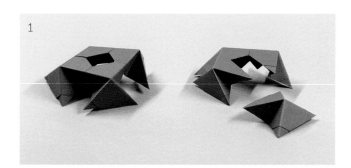

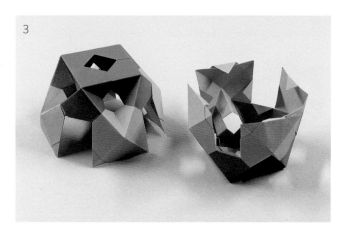

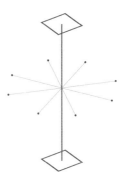

18-Side

First Steps
You will need 16 orange squares (8 x horse and 8 x rider) as well as 16 yellow squares (8 x horse and 8 x rider). Execute mountain and valley folds according to the illustrations on the right. Join horse and rider to form 8 orange and 8 yellow modules.

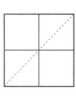

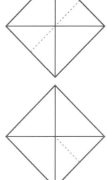

Joining the Model
Join 2 x 4 orange modules to create 2 square ring joints (fig. 1). 4 yellow modules each continue the orange modules (fig. 2), creating two identical halves of the model (fig. 3). Join the two halves together (joining yellow modules with yellow modules; fig. 4).

Similar Model
Without the yellow modules, you can use a similar process to create a cuboctahedron* (fig. 5).

Difficulty: Easy
Simple cutting, simple construction; time required: approximately 45 minutes.

*A02 in *Folding Polyhedra*

8 x   8 x   8 x   8 x

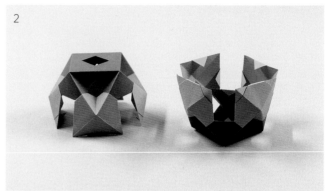

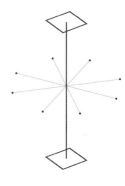

22-Side

First Steps

You will need 24 orange squares (12 x horse and 12 x rider) as well as 16 yellow squares (8 x horse and 8 x rider). Execute mountain and valley folds according to the illustrations on the right. Join horse and rider to form 12 orange and 8 yellow modules.

Joining the Model

Join 2 x 4 orange modules to form 2 square ring joints—just as in model O01. Join 4 yellow modules to each of the orange modules (fig. 1). This creates two identical parts of the model (fig. 2). Use 4 orange modules to join the two parts, creating the final form (figs. 3 and 4).

Similar Models

Obvious similarities exist between the cuboctahedron*, O02 and O01 (fig. 5, *from left to right*).

Difficulty: Easy

Simple cutting, simple construction; time required: approximately one hour.

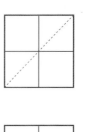

*A02 in *Folding Polyhedra*

12 x   12 x   8 x   8 x

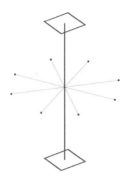

26-Side

First Steps
You will need 32 orange squares (16 x horse and 16 x rider) as well
as 16 yellow squares (8 x horse and 8 x rider). Execute mountain
and valley folds according to the illustrations on the right. Join
horse and rider to form 16 orange and 8 yellow modules.

Joining the Model
Join 2 x 4 orange modules to form 2 square ring joints, as before
in models O01 and O02. Continue with 4 yellow modules for
each orange module (fig. 1), then with 4 orange modules each
(fig. 2). This results in 2 identical parts of the model (fig. 3).
Join the two halves to create the final form (fig. 4).

Similar Models
Fig. 5 shows O01, O02, and O03 next to each other.

Difficulty: Easy
Simple cutting, simple construction; time required: approximately
1½ hours.

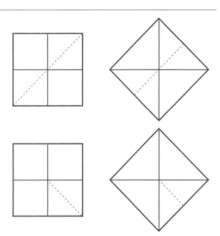

16 x   16 x   8 x   8 x

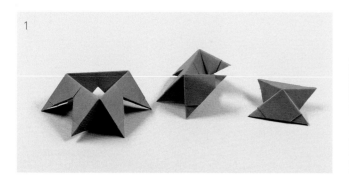

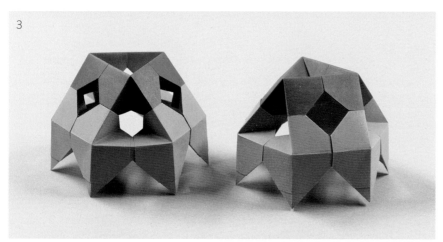

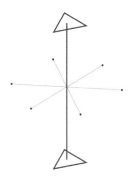

20-Side
(Two Versions)

First Steps

You will need 12 orange squares (6 x horse and 6 x rider) as well as 24 yellow squares (12 x horse and 12 x rider). Execute mountain and valley folds according to the illustrations on the right. Join horse and rider to form 5 orange and 12 yellow modules.

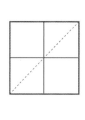
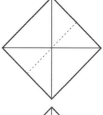

Joining the Model

Join 2 x 3 orange modules to form 2 triangular ring joints (fig. 1). Continue with 6 yellow modules each (fig. 2). These are 2 identical parts of the model (fig. 3). Join the two halves to form a closed shape (figs. 4 and 5).

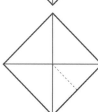

Difficulty: Easy

Simple cutting, simple construction; time required: approximately one hour.

6 x   6 x   12 x   12 x

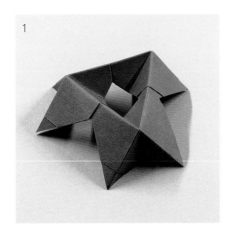

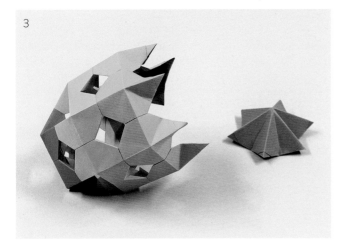

19-Side

First Steps

You will need 6 orange squares (3 x horse and 3 x rider) as well as 12 yellow squares (6 x horse and 6 x rider). Additionally, you will need a further 12 light-green squares (6 x horse and 6 x rider) as well as 2 moss-green hexagons (1 x horse and 1 x rider). Execute mountain and valley folds according to the illustrations on the right. Join horse and rider to form 3 orange, 6 yellow, 6 light-green, and one moss-green module.

Joining the Model

Join 3 orange modules to form a triangular ring joint (fig. 1). Continue from the orange modules with 6 yellow modules (fig. 2). Then join 6 light-green modules to these, in a similar way as in model N03 (fig. 3). The six light-green tips are topped with a concluding moss-green module (fig. 4). For this model, you can access video instructions by using this link: https://www.haupt.ch/faltformen.

Similar Models

As can be seen in fig. 5, O05 is a combination of N03 and O04.

Difficulty: Moderate

Cutting: some challenges, moderately challenging construction; time required: approximately one hour.

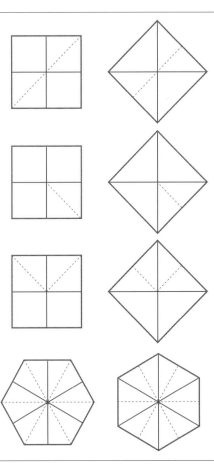

3 x   3 x   6 x   6 x   6 x   6 x   1 x   1 x

23-Side

First Steps

You will need 6 orange squares (3 x horse and 3 x rider) as well as 24 yellow squares (12 x horse and 12 x rider). Execute mountain and valley folds according to the illustrations on the right. Join horse and rider to form 3 orange and 12 yellow modules.

Joining the Model

Join 2 x 3 yellow modules to form two triangular ring joints (fig. 1); glue and clips will help keep the joints in place. Continue with 6 further yellow modules each (fig. 2). This results in two identical parts consisting of 9 modules each. Add one orange module to the first three tips (fig. 3). Then add the second part like a lid to the first. Fig. 4 shows both parts before they are joined together (*left, right*); the completed model is in the center (see also fig. 5).

Difficulty: Moderate

Simple cutting, moderately difficult construction; time required: approximately 1½ hours.

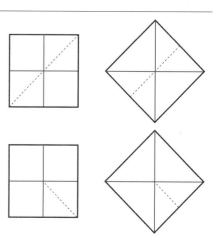

Note:
When joining adjacent modules, use glue and fix modules in place with clips until the glue has dried completely.

3 x	3 x	12 x	12 x

1

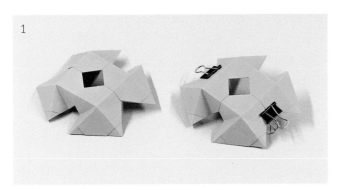

2

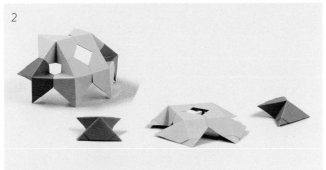

3

4

5

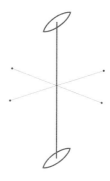

22-Side

First Steps
You will need 8 orange squares (4 x horse and 4 x rider) as well as 32 yellow squares (16 x horse and 16 x rider). Execute mountain and valley folds according to the illustrations on the right. Join horse and rider to form 4 orange and 16 yellow modules.

Joining the Model
Join 2 x 4 yellow modules to form a square ring joint each (fig. 1). Add 2 orange modules at two sides each (fig. 2). Add 2 yellow modules at the other sides (fig. 3). This results in 2 identical parts of the model. Fig. 4 shows both halves (*left and right*); in the center, the completed model (see also fig. 5).

Difficulty: Moderate
Simple cutting, moderately difficult construction; time required: approximately 1½ hours.

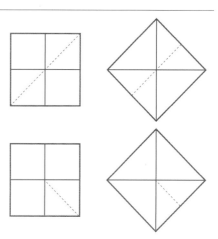

Note:
When joining adjacent modules, use glue and fix modules in place with clips until the glue has dried completely.

4 x   4 x   16 x   16 x

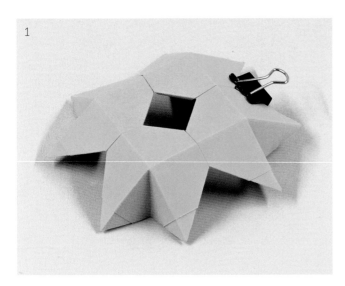

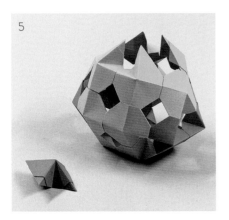

21-Side

First Steps

You will need 10 orange squares (5 x horse and 5 x rider) as well as 28 yellow squares (14 x horse and 14 x rider). Execute mountain and valley folds according to the illustration on the right. Join horse and rider to form 5 orange and 14 yellow modules.

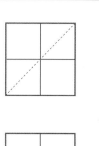
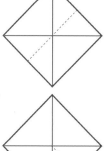

Joining the Model

Join 4 yellow modules to form a square ring joint (fig. 1). 2 orange modules continue this on two sides (fig. 2). Using a yellow module, create triangular ring joints with the orange modules (fig. 3). This creates a bowl-like form, as illustrated in fig. 4. Add the remaining 2 x 2 yellow modules at the sides (fig. 4). The last orange module acts as a cornerstone (fig. 5). Fig. 6 shows the completed model.

Difficulty: Moderate

Simple cutting, moderately difficult construction; time required: approximately 1½ hours.

Hint:
When plugging adjacent modules together, use glue and lock the modules with staples until the adhesive has dried.

5 x	5 x	14 x	14 x

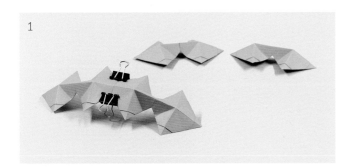

1

2

3

4

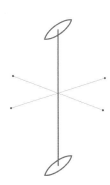

18-Side (Chiral)

First Steps
You will need 16 light green squares (8 x horse and 8 x rider) as well as 16 yellow squares (8 x horse and 8 x rider). Execute mountain and valley folds according to the illustration on the right. Then join horse and rider to form 8 light-green and 8 yellow modules.

Joining the Model
Join 2 x 4 light green modules to form a row (fig. 1). Add a yellow module to each end (fig. 2). This creates two identical parts of the model. Join these two parts together (fig. 3). This can be done in one of two possible (chiral) ways—join the two halves in such a way that the two light-green rows do not touch directly. The two models are mirror symmetrical to each other (fig. 4).

Similar Models
Fig. 5 shows that this form is related to that of the rhombic cuboctahedron top* (*in the center of the picture*). Here, the light-green modules run once around the entire model.

Difficulty: Moderate
Simple cutting, moderately difficult construction; time required: approximately 1½ hours.

*J03 from *Folding Polyhedra*

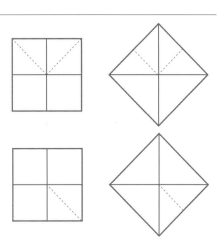

Note:
When joining adjacent modules, use glue and fix modules in place with clips until the glue has dried completely.

8 x	8 x	8 x	8 x

P Models:
From Squares
(Mixed)

The series of P models continues the combination of square modules and increases the range by adding dark-red modules (3 valley folds). P01 is a singular form with a fourfold main symmetry axis. P02 to P04 form their own construction series exclusively with threefold symmetry axes.

P05 and P06 are peculiar forms with only one twofold symmetry axis. This is also the case with P07, but this form has an additional two 2-fold subsidiary symmetry axes. P08 is a chiral form and can be constructed in two mirroring versions.

The attraction of the P models may lie less in their first aesthetic impression. It certainly lies in the subtle possibilities of combining modules, as well as in the special ratios of symmetries, which are surprisingly difficult to decipher in some of these models.

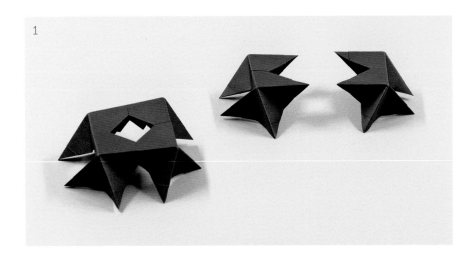

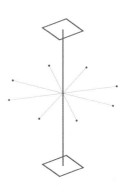

10-Side

First Steps

You will need 16 dark red squares (8 x horse and 8 x rider). Execute mountain and valley folds according to the illustration on the right. Join horse and rider to form 8 dark-red modules.

Joining the Model

Join 2 x 4 modules to form 2 square ring joints (fig. 1). This already creates the two halves of the model. Fig. 2 shows both halves (*left, right*) which are joined together to form a complete model (*front/central in picture*). Fig. 3 shows the model from different perspectives.

Difficulty: Moderate

Simple cutting, tricky construction; time required: approximately 45 minutes.

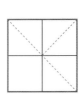
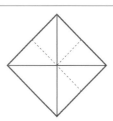

8 x   8 x

11-Side

First Steps

You will need 12 dark-red squares (6 x horse and 6 x rider) as well as 6 orange squares (3 x horse and 3 x rider). Execute mountain and valley folds according to the illustrations on the right. Join horse and rider to form 6 dark-red modules and 3 orange modules.

Joining the Model

Join 2 x 3 dark-red modules to form a triangular ring joint each (fig. 1). These will serve as the top and bottom ends of the model. Both are joined together using the three orange modules (fig. 2). Fig. 3 shows P01 with a 4-fold symmetry next to P02 with a 3-fold symmetry.

Difficulty: Moderate

Simple cutting, tricky construction; time required: approximately one hour.

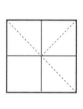
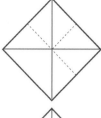

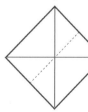

6 x	6 x	3 x	3 x

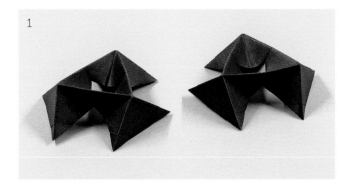

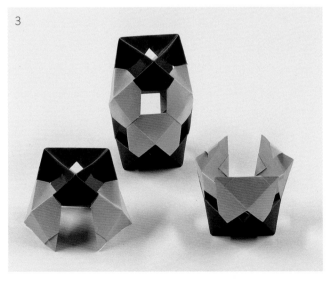

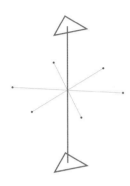

14-Side

First Steps

You will need 12 dark-red squares (6 x horse and 6 x rider) as well as 12 yellow squares (6 x horse and 6 x rider). Execute mountain and valley folds according to the illustration on the right. Join horse and rider to form 6 dark-red and 6 yellow modules. It is helpful to glue the folded-over tips of the horses on the underside (the smooth side) of the modules, as seen in the picture.

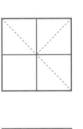

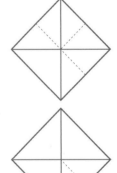

Joining the Model

Join 2 x 3 modules to form a triangular ring joint each. These will serve as the top and bottom end of the model (fig. 1, as in model P02). 3 yellow modules each are joined to these (fig. 2). This creates the two sides of the model (fig. 3, *left, right*). Join these to create the finished model (fig. 3, *center/back*).

Similar Models

Fig. 4 shows the obvious relationship between the forms of P02 and P03.

Difficulty: Moderate

Simple cutting, tricky construction; time required: approximately one hour.

6 x   6 x   6 x   6 x

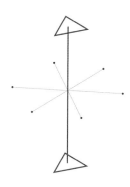

17-Side

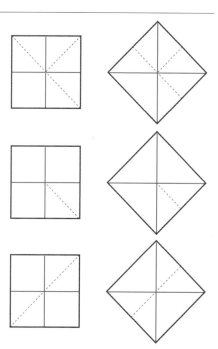

First Steps

You will need 12 dark-red squares (6 x horse and 6 x rider), 12 yellow squares (6 x horse and 6 x rider), and 6 orange squares (3 x horse and 3 x rider). Execute mountain and valley folds according to the illustrations on the right. Join horse and rider to form 6 dark-red, 6 yellow, and 3 orange modules. It is helpful to secure the folded-over tips of the rider parts with a small amount of glue on the underside (the smooth side) of each module (fig. 1).

Joining the Model

Join 2 x 3 dark-red modules to form a triangular ring joint each. These will form the upper and lower end of the model (as in models P02 and P03, here without picture). 3 yellow modules are joined to each (fig. 1); this creates two identical parts of the model. These are eventually joined using 3 orange modules (fig. 2, *left, right*; the completed model is shown centrally, at the back).

Similar Models

Figs. 3 and 4 show the obvious similarities in the forms of P02, P03, and P04 from different perspectives.

Difficulty: Moderate

Simple cutting, tricky construction; time required: approximately one hour.

| 6 x | 6 x | 6 x | 6 x | 3 x | 3 x |

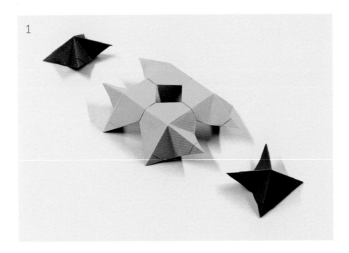

13-Side

First Steps
You will need 10 orange squares (5 x horse and 5 x rider), 4 dark-red squares (2 x horse and 2 x rider), and 8 light-green squares (4 x horse and 4 x rider). Execute mountain and valley folds according to the illustrations on the right. Join horse and rider to form 5 orange, 2 dark-red, and 4 light-green modules.

Joining the Model
Join 4 light-green modules to form a square ring joint (fig. 1). Continue at 2 sides with a dark-red module each (fig. 2), and at the remaining open light-green tips with 2 orange modules each (fig. 3). A further orange module serves as the completing stone (fig. 4). Fig. 5 shows the completed model from 2 different perspectives.

Similar Models
O08, P06, and P05 have the shared characteristic of a 2-fold main symmetry axis.

Difficulty: Easy
Simple cutting, simple construction; time required: approximately 45 minutes.

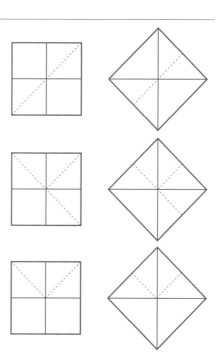

5 x	5 x	2 x	2 x	4 x	4 x

1

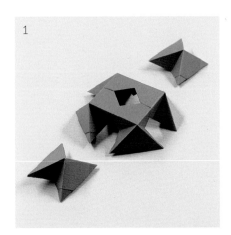

2

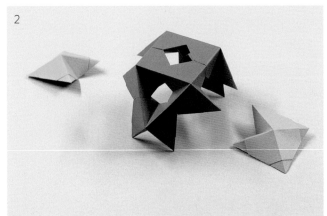

3

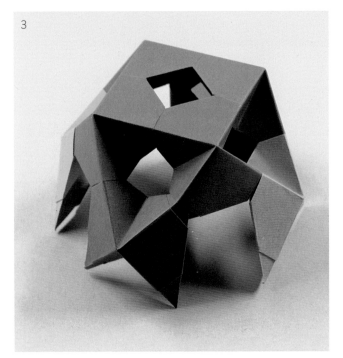

4

6

5

7

15-Side

First Steps
You will need 14 orange squares (7 x horse and 7 x rider), 4 yellow squares (2 x horse and 2 x rider), and 8 light-green squares (4 x horse and 4 x rider). Execute mountain and valley folds according to the illustrations on the right. Join horse and rider to form 7 orange, 2 yellow, and 8 light-green modules.

Joining the Model
Join 4 orange modules to form a square ring joint (fig. 1). Continue with an orange module each on 2 sides (fig. 2), at the remaining open orange tips, with a yellow module each (fig. 3). 4 light-green modules are added (fig. 4). An orange module completes the model (fig. 5). Fig. 6 shows the model from 2 different perspectives.

Similar Models
Fig. 7 shows three models with a 2-fold main symmetry axis: O08, P06, and P05 (*from left to right*).

Difficulty: Easy
Simple cutting, simple construction; time required: approximately 45 minutes.

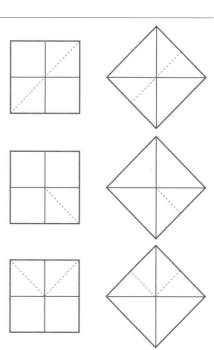

| | 7 x | 7 x | 2 x | 2 x | 4 x | 4 x |

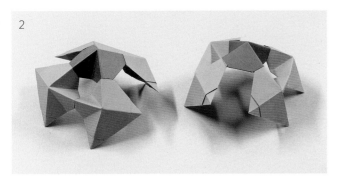

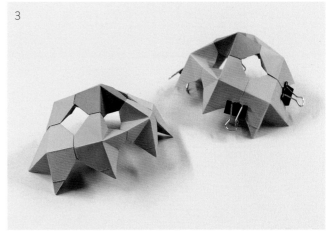

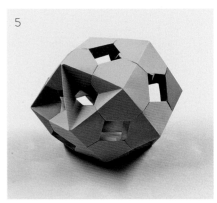

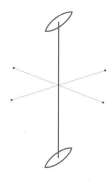

16-Side

First Steps

You will need 4 orange squares (2 x horse and 2 x rider), 8 yellow squares (4 x horse and 4 x rider), and 16 light-green squares (8 x horse and 8 x rider). Execute mountain and valley folds according to the illustrations on the right. Join horse and rider to form 2 orange, 4 yellow, and 8 light-green modules.

Joining the Model

Join 4 light-green modules to each of the two orange modules (figs. 1 and 2). This creates two identical parts of the model, which are joined together using 4 yellow modules (figs. 3 and 4). Fig. 5 shows the completed model.

Difficulty: Easy

Simple cutting, simple construction; time required: approximately 45 minutes.

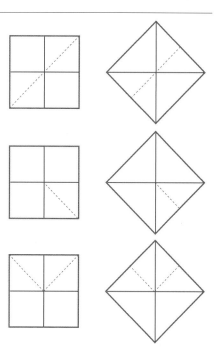

Hint:

It is helpful to support the plug connections with adhesive.

2 x	2 x	4 x	4 x	8 x	8 x

1

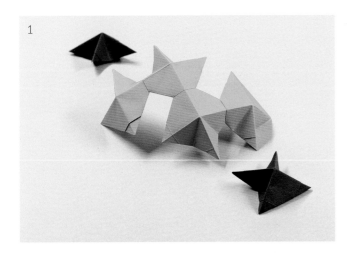

2

3

4

5

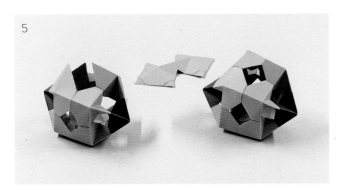

6

14-Side (Chiral)

First Steps
You will need 4 orange squares (2 x horse and 2 x rider), as well as 4 dark-red squares (2 x horse and 2 x rider). Additionally, you will need a further 4 yellow squares (2 x horse and 2 x rider) as well as 12 light-green squares (6 x horse and 6 x rider). Execute mountain and valley folds according to the illustrations on the right. Join horse and rider to form 2 orange, 2 dark-red, 2 yellow, and 6 light-green modules.

Joining the Model
Join 4 light-green modules to create a stair-like row (fig. 1, central). This is continued with dark-red (fig. 1, outside) and yellow modules (figs. 2 and 3). 2 orange modules follow (fig. 4). 2 joined light-green modules complete the model (fig. 5). Two mirror symmetrical solutions are possible (fig. 6).

Similar Models
O09 and P08 can be realized in two chiral versions each (fig. 6).

Difficulty: Moderate
Simple cutting, tricky construction; time required: approximately one hour for each version.

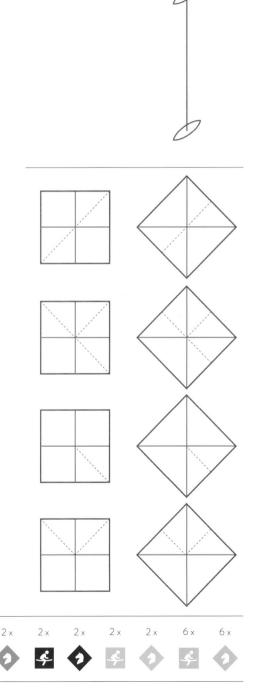

| 2 x | 2 x | 2 x | 2 x | 2 x | 2 x | 6 x | 6 x |

Q Models:
From Triangles
(Pink)

With the Q models, we're using triangular models for the first time, and we're using pink modules nearly exclusively. These models are very similar to the highly symmetrical forms of the truncated tetrahedron, truncated octahedron, and truncated icosahedron*.

Unlike their relatives, however, the Q models are very much reduced to the principle of one main symmetry axis with several subsidiary symmetry axes. Many solutions of parts of Q models can be seen in models of the S and T series in a slightly modified form.

*B05, B06, B07, and the L models from *Folding Polyhedra*

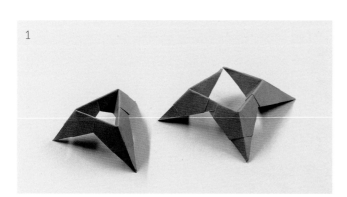

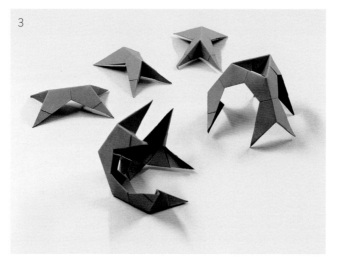

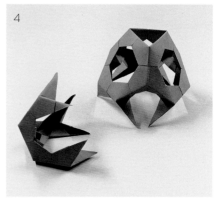

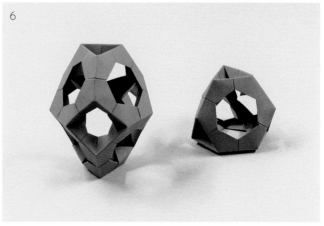

11-Side

First Steps

You will need 36 pink triangles (18 x horse and 18 x rider). Execute mountain and valley folds according to the illustration on the right. Join horse and rider to form 18 pink modules.

Joining the Model

Join 2 x 3 modules to form a triangular ring joint (fig. 1). Continue with 1 module each (figs. 2 and 3). This results in 2 identical parts of the model. Complete the model using the remaining 3 x 2 modules (fig. 3, *above*). Figs. 4 and 5 show the completion process.

Similar Models

It is obvious that Q01 is similar to the truncated tetrahedron*, as fig. 6 shows.

Difficulty: Easy

Simple cutting, simple construction; time required: approximately 45 minutes.

18 x   18 x

*B05 from *Folding Polyhedra*.

1
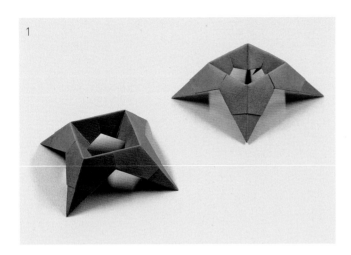

2

3
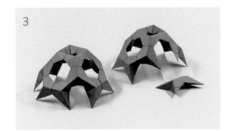

4
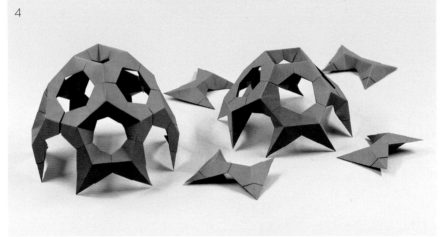

5

6

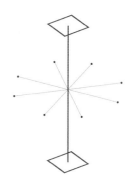

26-Side

First Steps

You will need 96 pink triangles (48 x horse and 48 x rider). Execute mountain and valley folds according to the illustration on the right. Join horse and rider to form 48 pink modules.

Joining the Model

Join 2 x 4 modules to form a square ring joint (fig. 1). One module is then added to each tip (fig. 2). This is followed by 4 double modules each (fig. 3), and a further 4 double modules (fig. 4). This results in two identical parts of the model (fig. 5), which are joined in the last step (fig. 6).

Difficulty: Easy

Simple cutting, simple construction; time required: approximately two hours.

48 x   48 x

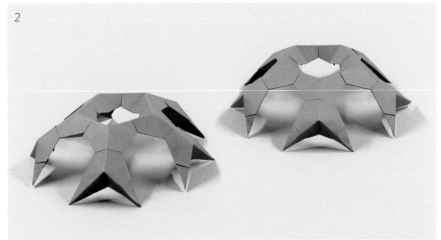

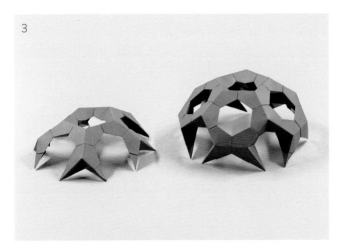

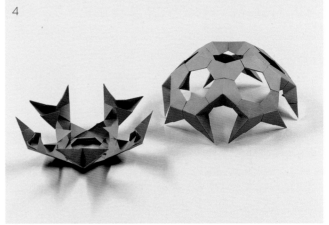

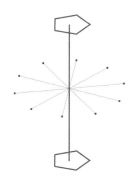

17-Side

First Steps
You will need 60 pink triangles (30 x horse and 30 x rider).
Execute mountain and valley folds according to the illustration
on the right. Join horse and rider to form 30 modules.

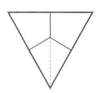

Joining the Model
Join 2 x 5 modules to create a pentagonal ring joint (fig. 1).
Continue with one module each (fig. 2). This results in 2 identical
parts of the model, which are joined using 5 x 2 modules (figs.
3 and 4). This creates the final model (fig. 5).

Similar Models
The models Q01, the truncated octahedron*, and Q03 form
their own series. They all have radial 4-fold ring joints; the main
axes are 3-, 4-, and 5-fold**. It is very obvious that Q03 also
shares similarities with the truncated icosahedron*** (see model
Q05, fig. 6, *in the background*).

Difficulty: Easy
Simple cutting, simple construction; time required: approximately
two hours.

*B06 from *Folding Polyhedra*. / **B06, however, has not only one main symmetry axis but a much more complex symmetry structure than Q01/Q03 (fig. 6).
/ ***B07 from *Folding Polyhedra*.

30 x   30 x

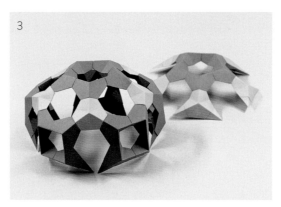

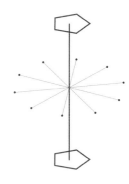

22-Side

First Steps
You will need 60 pink triangles (30 x horse and 30 x rider) as well as 20 light-blue triangles (10 x horse and 10 x rider). Execute mountain and valley folds according to the illustrations on the right. Join horse and rider to form 30 pink and 10 light-blue modules.

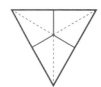

Assembling the Model
2 x 5 pink modules are joined to form a pentagonal ring joint each, to which a light-blue module is added (fig. 1). This creates two identical parts of the model. Join the remaining 4 x 5 pink modules to form square ring joints (fig. 2); these serve to connect the two parts constructed before (fig. 3).

Alternative Construction
Q04 can also be built using a single color (fig. 4).

Difficulty: Easy
Simple cutting, simple construction; time required: approximately two hours.

30 x	30 x	10 x	10 x

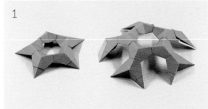

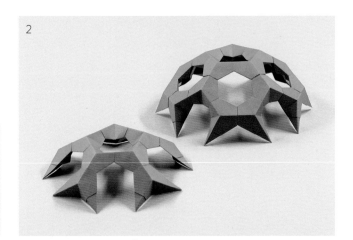

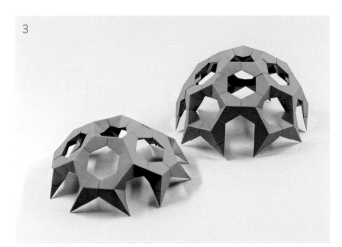

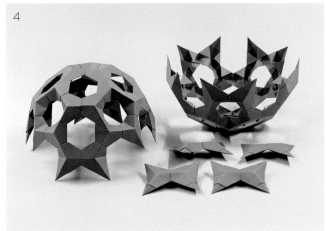

37-Side
(Twin Truncated
Icosahedron)

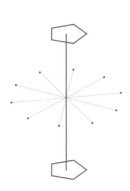

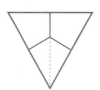
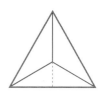

First Steps

You will need 140 pink triangles (70 x horse and 70 x rider). Execute mountain and valley folds according to the illustration on the right. Join horse and rider to form 70 pink modules.

Assembling the Model

2 x 5 modules are joined to form a pentagonal ring joint each (fig. 1, *left*). These are continued with one module each (fig. 1, *right*, and fig. 2, *left*). 5 double modules follow for each part (fig. 2, *right*) and a further 5 double modules (fig. 3, *right*). This creates two identical halves of the model (fig. 4), which are connected in a final step using a further 5 double modules (fig. 4; see also model Q04).

Similar Models

Fig. 5 shows the completed Q05 together with the highly symmetrical truncated icosahedron*.

Difficulty: Easy

Simple cutting, simple construction; time required: approximately 2½ hours.

*B07 from *Folding Polyhedra*.

70 x   70 x

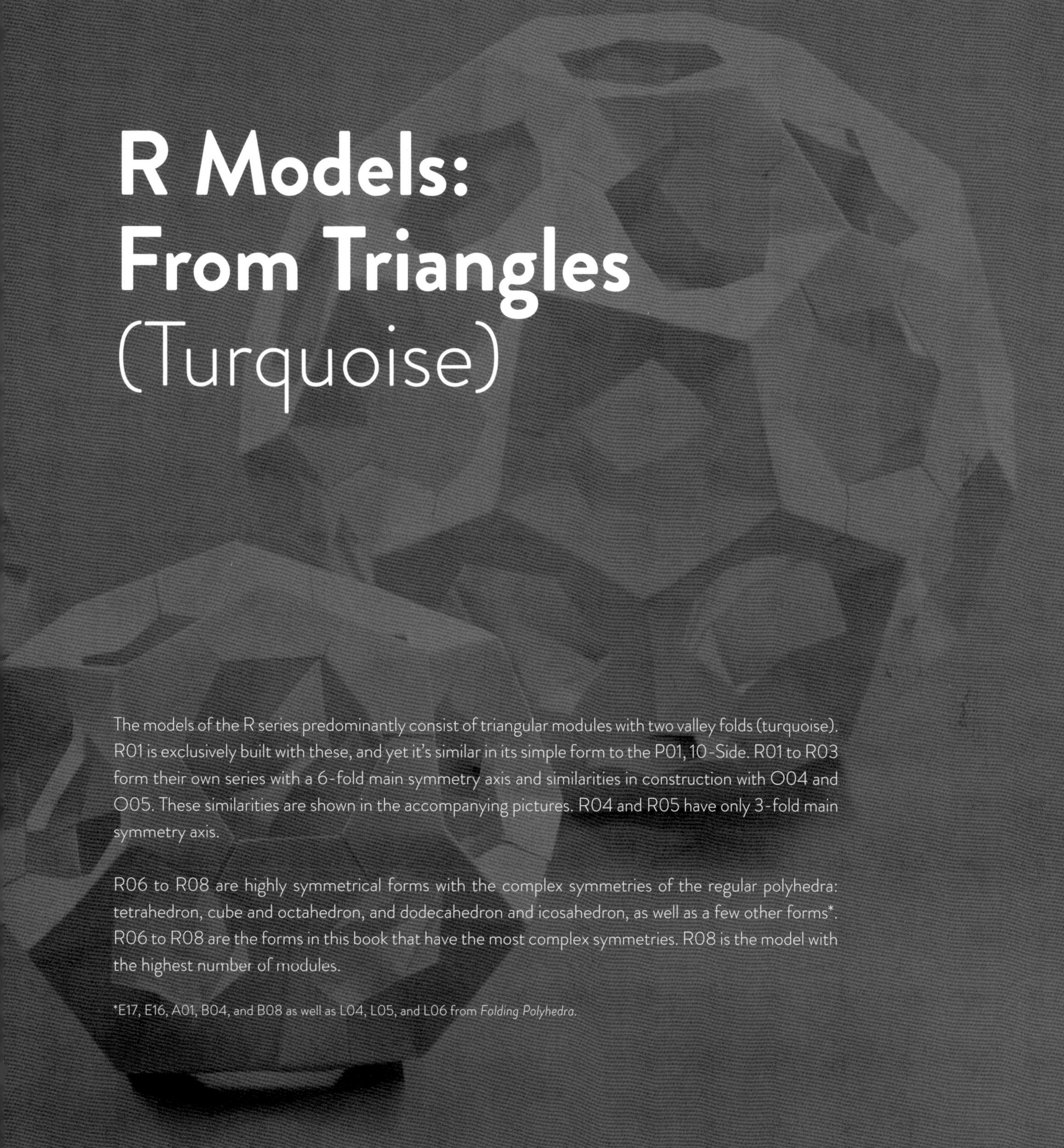

R Models:
From Triangles
(Turquoise)

The models of the R series predominantly consist of triangular modules with two valley folds (turquoise). R01 is exclusively built with these, and yet it's similar in its simple form to the P01, 10-Side. R01 to R03 form their own series with a 6-fold main symmetry axis and similarities in construction with O04 and O05. These similarities are shown in the accompanying pictures. R04 and R05 have only 3-fold main symmetry axis.

R06 to R08 are highly symmetrical forms with the complex symmetries of the regular polyhedra: tetrahedron, cube and octahedron, and dodecahedron and icosahedron, as well as a few other forms*. R06 to R08 are the forms in this book that have the most complex symmetries. R08 is the model with the highest number of modules.

*E17, E16, A01, B04, and B08 as well as L04, L05, and L06 from *Folding Polyhedra*.

1

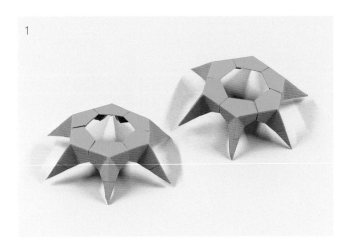

2

3

4

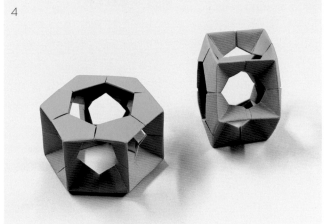

5

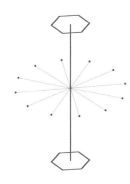

8-Side

First Steps
You will need 24 turquoise triangles (12 x horse and 12 x rider).
Execute mountain and valley folds according to the illustration
on the right. Join horse and rider to form 12 turquoise modules.

Assembling the Model
Join 2 x 6 modules to form a hexagonal ring joint each (figs. 1
and 2). Then join the two parts (fig. 3). This requires some effort
and presents some difficulties. Fig. 4 shows the completed
model from different perspectives.

Similar Models
R01 is the starting point for the similar models R02 and R03
(fig. 5).

Difficulty: Easy to Medium
Simple cutting, moderately difficult construction; time required:
approximately 45 minutes.

12 x   12 x

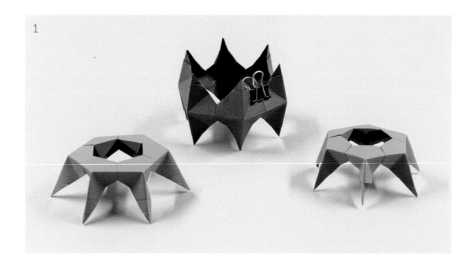

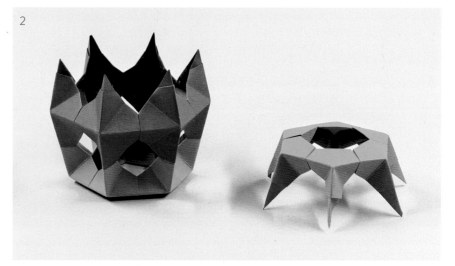

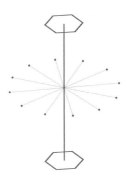

14-Side

First Steps

You will need 24 turquoise triangles (12 x horse and 12 x rider) as well as 12 dark-blue squares (6 x horse and 6 x rider). Execute mountain and valley folds according to the illustrations on the right. Join horse and rider to form 12 turquoise and 6 dark-blue modules.

 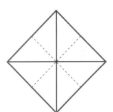

Assembling the Model

Join 2 x 6 turquoise modules to form a hexagonal ring joint each (fig. 1). This creates two identical parts of the model. Join the 6 dark-blue modules to form a tubular series. Use glue to support the joints (fig. 1, *center*). Attach the dark-blue tube to the rim of the first 6-module part (fig. 2). Attach the second 6-module part from the other side (fig. 3). This is somewhat tricky. Fig. 4 shows the completed model from a different perspective.

Difficulty: Easy to Moderate

Easy cutting, moderately difficult construction; time required: approximately one hour.

12 x   12 x   6 x   6 x

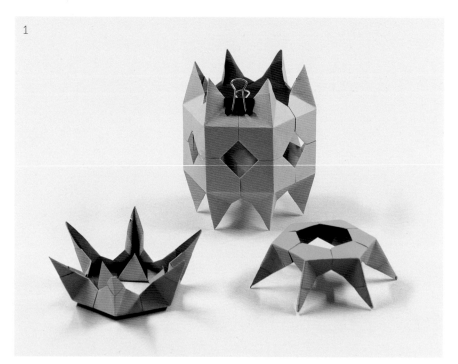

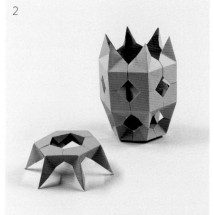

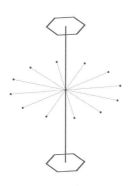

20-Side

First Steps

You will need 24 turquoise triangles (12 x horse and 12 x rider) as well as 24 light-green squares (12 x horse and 12 x rider). Execute mountain and valley folds according to the illustrations on the right. Join horse and rider to form 12 turquoise and 12 light-green modules.

Assembling the Model

Join 2 x 6 turquoise modules to form a hexagonal ring joint each (fig. 1, *right, left*). This creates two identical turquoise parts of the model. Then join the 12 light-green modules to form a kind of tube (fig. 1, *center back*). The light-green tube serves as the link between the two turquoise parts (figs. 2 and 3). Joining the parts is somewhat tricky.

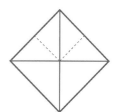

Similar Models

Partial solutions of the similar models N03 and O06 can be transferred to one-half of R03.

Difficulty: Easy to Moderate

Simple cutting, simple to moderately difficult construction; time required: approximately one hour.

12 x   12 x   12 x   12 x

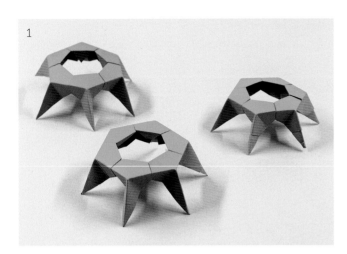

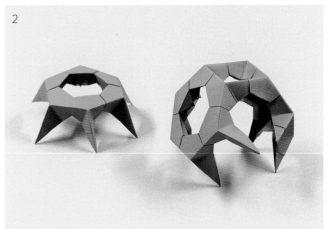

12-Side

First Steps

You will need 36 turquoise triangles (18 x horse and 18 x rider) as well as 4 light-blue triangles (2 x horse and 2 x rider). Execute mountain and valley folds according to the illustrations on the right. Join horse and rider to form 18 turquoise and 2 light- blue modules.

Assembling the Model

Join 2 x 6 turquoise modules to form 3 hexagonal ring joints (fig. 1). 2 ring joints are joined with each other at 2 tips (fig. 2). The third ring joint is added to the first two in the same way, resulting in a tubular structure (fig. 3). A light-blue module forms the capstone at each end. Figs. 4 and 5 show the model from different perspectives.

Difficulty: Easy

Simple cutting, simple construction; time required: approximately one hour.

18 x   18 x   2 x   2 x

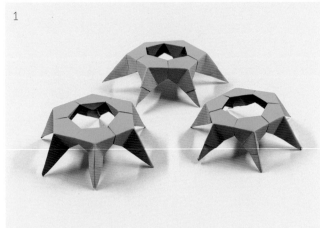

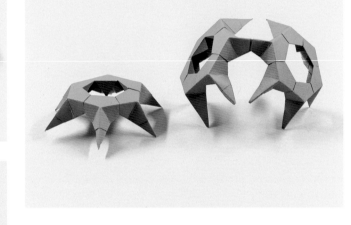

15-Side

First Steps
You will need 36 turquoise triangles (18 x horse and 18 x rider) as well as 16 light-blue triangles (8 x horse and 8 x rider). Execute mountain and valley folds according to the illustrations on the right. Then join horse and rider to form 18 turquoise and 8 light-blue modules.

Assembling the Model
Join 3 x 6 turquoise modules to form 3 hexagonal ring joints (fig. 1) and then join these only at one tip, respectively (fig. 2). This creates a tubular structure (fig. 3). The light-blue modules are assembled in groups of four (fig. 3). These complete the model at 2 ends as capstones consisting of four parts. Figs. 4 and 5 show the model from different perspectives.

Difficulty: Easy
Simple cutting, simple construction; time required: approximately 1¼ hours.

18 x   18 x   8 x   8 x

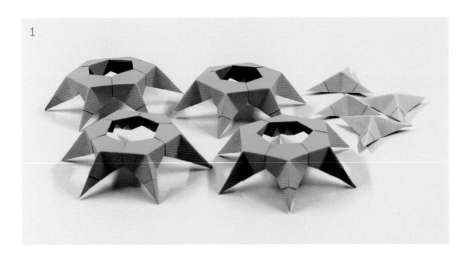

16-Side

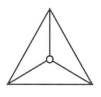

First Steps

You will need 48 turquoise triangles (24 x horse and 24 x rider) as well as 8 light-blue triangles (4 x horse and 4 x rider). Execute mountain and valley folds according to the illustrations on the right. Then join horse and rider to form 24 turquoise and 4 light-blue modules.

Assembling the Model

4 x 6 modules are joined to form 4 hexagonal ring joints (fig. 1). Join the first two turquoise ring joints with each other at one tip, and additionally with two light-blue modules (fig. 2). The third ring joint is added in the same way, as are the remaining 2 light-blue modules (fig. 3). The last ring joint completes the model. Figs. 4 and 5 show the completed model from different perspectives.

Difficulty: Easy

Simple cutting, simple construction; time required: approximately 1¼ hours.

24 x   24 x   4 x   4 x

1

2

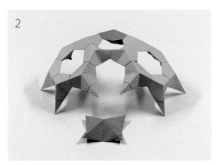

3

4

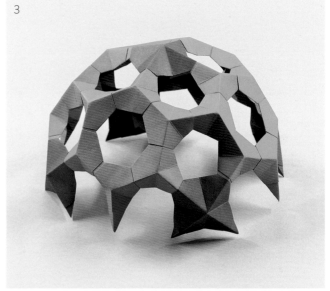

5

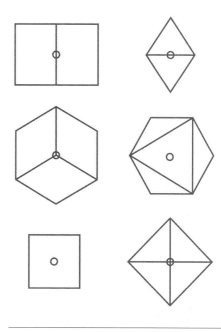

32-Side°

First Steps

You will need 96 turquoise triangles (48 x horse and 48 x rider) as well as 12 dark-blue squares (6 x horse and 6 x rider). Execute mountain and valley folds according to the illustrations on the right. Join horse and rider to form 48 turquoise and 12 dark-blue modules.

Assembling the Model

Join 8 x 6 turquoise modules to form 8 hexagonal ring joints (fig. 1). The first two are joined with each other and additionally with 2 dark-blue modules (fig. 2). Continue according to the same pattern (figs. 3 and 4): join dark-blue modules only with turquoise ring joints. Turquoise ring joints are alternatingly joined with other turquoise ring joints and dark-blue modules. Fig. 5 shows the finished model.

Difficulty: Easy

Simple cutting, simple construction; time required: approximately 1¼ hours.

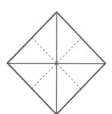

Note:

The blue squares in this project are larger than the standard size (see page 18). Horse and rider in deviating sizes (see https://www. haupt.ch/faltformen).
°Some paper sizes used in this project deviate from the standard measurements (see page 18).

	48 x	48 x	6 x	6 x

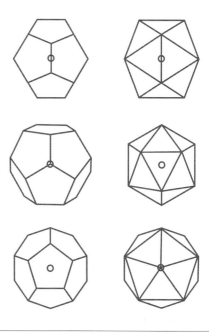

80-Side°

First Steps

You will need 240 turquoise triangles (120 x horse and 120 x rider) as well as 24 violet pentagons (12 x horse and 12 x rider). Execute mountain and valley folds according to the illustrations on the right. Then join horse and rider to form 120 turquoise and 12 violet modules.

Assembling the Model

Join 20 x 6 turquoise modules to form 20 hexagonal ring joints (as in R06, R07; without illustration here) and join these according to the same pattern (fig. 1). Each hexagonal turquoise part has only one joint with each adjacent identical turquoise part, every second possible joint of the hexagonal parts stays open for the moment. Continue according to the same pattern (figs. 2–5). Fig. 6 shows the model before the last gaps are closed with a violet capstone. Fig. 7 shows the completed model.

Difficulty: Easy to Moderate

Cutting: some challenges, simple construction; time required: approximately four hours.

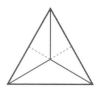

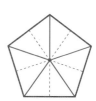

Note:

The violet pentagons are bigger than the standard size (see page 18).

Horse and rider in deviating sizes (see https://www.haupt.ch/faltformen).

° Some paper sizes used in this model deviate from the standard size (see page 18).

120 x	120 x	12 x	12 x

S Models:
From Triangles
(Mixed)

The models of the S series share many topological similarities with those of the Q series. Their construction is mainly based on pink and turquoise modules. S01 to S03 form their own series, using the same types of modules, while the number of symmetries of their main symmetry axis increases progressively.

In a similar way, this also happens with models S04 to S06. These have alternative solutions at two ends of each model. S07 and S08 exhibit further 3- and 4-fold solutions, respectively. S09 and S10 have only one 2-fold main symmetry axis and two 2-fold subsidiary symmetry axes each.

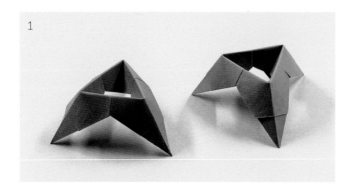

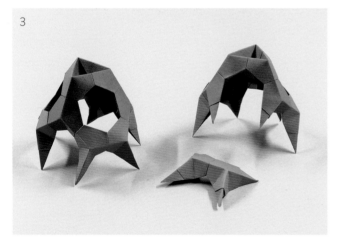

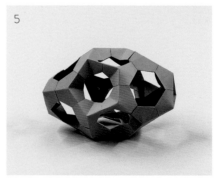

14-Side

First Steps

You will need 24 pink triangles (12 x horse and 12 x rider) as well as 24 turquoise triangles (12 x horse and 12 x rider). Execute mountain and valley folds according to the illustrations on the right. Join horse and rider to form 12 pink and 12 turquoise modules.

Assembling the Model

Join 2 x 3 pink modules to form 2 triangular ring joints (fig. 1). Each open joint is followed by another pink module (fig. 2). 2 turquoise modules follow at each end (fig. 3). This results in two identical parts of the model (fig. 4), which only have to be put together (fig. 5). If you find this difficult to accomplish, start with model S02.

Difficulty: Easy to Moderate

Simple cutting, slightly tricky construction; time required: approximately 45 minutes.

12 x   12 x   12 x   12 x

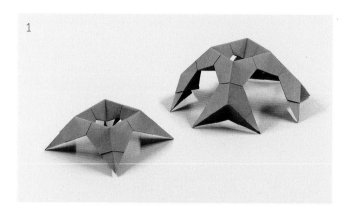

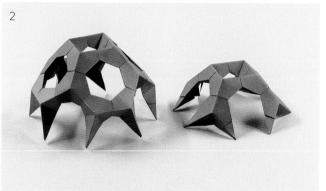

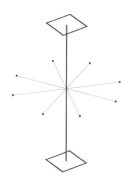

18-Side

First Steps
You will need 32 pink triangles (16 x horse and 16 x rider) as well as 32 turquoise triangles (16 x horse and 16 x rider). Execute mountain and valley folds according to the illustrations on the right. Then join horse and rider to form 16 pink modules and 16 turquoise modules.

Assembling the Model
Join 2 x 4 pink modules to form 2 square ring joints (fig. 1, *left*). Each open joint is followed by a further pink module (fig. 1, *right*). At each end, 2 turquoise modules are added (fig. 2, *left*). This creates two identical parts of the model (fig. 3), which are now joined together to form the completed model (fig. 4).

Difficulty: Easy
Simple cutting, simple construction; time required: approximately one hour.

16 x   16 x   16 x   16 x

1

2

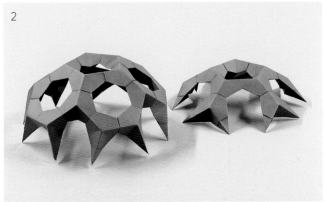

3

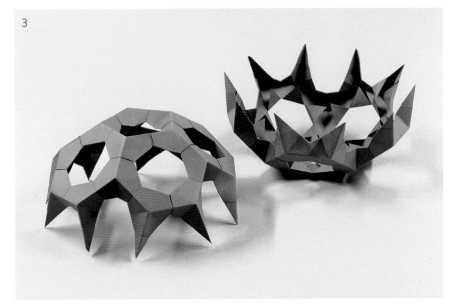

4

5

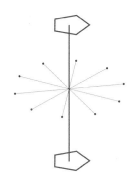

22-Side

First Steps
You will need 40 pink triangles (20 x horse and 20 x rider) as well as 40 turquoise triangles (20 x horse and 20 x rider). Execute mountain and valley folds according to the illustrations on the right. Then join horse and rider to form 20 pink and 20 turquoise modules.

Assembling the Model
Join 2 x 5 pink modules to form 2 pentagonal ring joints (fig. 1, *left*). Each open joint is followed by another pink module (fig. 1, *right*). 2 turquoise modules are added at each end (fig. 2, *left*). This results in two identical halves of the model (fig. 3), which are then joined to form the completed model (fig. 4).

Similarities
S01, S02, and S03 obviously belong together. The number of symmetries of the main axis increases from one model to the next (fig. 5).

Difficulty: Easy
Simple cutting, simple construction; time required: approximately 1½ hours.

	20 x	20 x	20 x	20 x

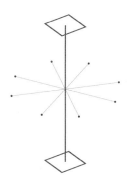

12-Side

First Steps
You will need 4 dark-blue squares (2 x horse and 2 x rider), 16 pink triangles (8 x horse and 8 x rider), and 16 turquoise triangles (8 x horse and 8 x rider). Execute mountain and valley folds according to the illustrations on the right. Join horse and rider to form 2 dark-blue, 8 pink, and 8 turquoise modules.

Assembling the Model
Join one dark-blue module with a turquoise module at each of the four ends (fig. 1). Repeat the same with the second dark-blue module. The turquoise modules are followed by a 4 x 1 pink module (fig. 2). This results in two parts that are joined to form the completed model (fig. 3).

Similarities
S04, S05, and S06 form a series in which the number of symmetries of the main symmetry axis increases step by step (fig. 4).

Difficulty: Easy
Simple cutting, simple construction; time required: approximately 45 minutes.

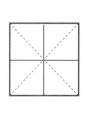
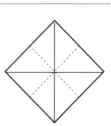

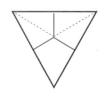
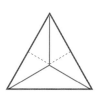

2 x	2 x	8 x	8 x	8 x	8 x

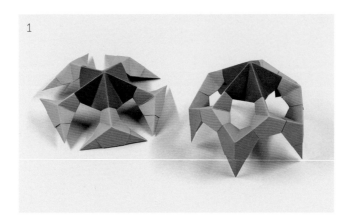

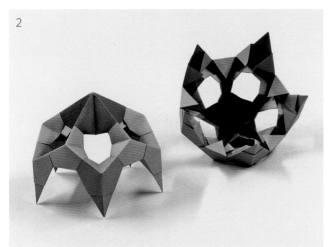

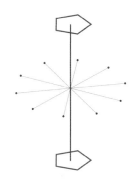

15-Side

First Steps
You will need 20 pink triangles (10 x horse and 10 x rider), 20 turquoise triangles (10 x horse and 10 x rider), and 4 violet pentagons (2 x horse and 2 x rider). Execute mountain and valley folds according to the illustrations on the right. Join horse and rider to form 10 pink, 10 turquoise, and 2 violet modules.

Assembling the Model
Join each of the two violet modules with one turquoise module at each of the five ends (fig. 1, *left*). These are followed by a 5 x 1 pink module (fig. 1, *right*). This results in two halves (fig. 2), which are joined to form the completed model (fig. 3).

Similarities
S05 (fig. 4, *center*) can be built with the following variations: instead of a violet pentagonal module, you can also use 5 light-blue triangular modules. Fig. 4 shows variations as 16-Side (*left*) and 17-Side (*right*), here without instructions.

Difficulty: Easy to Moderate
Simple to moderately difficult cutting, simple to moderately difficult construction; time required: approximately 1½ hours.

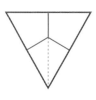

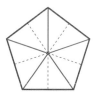

10 x	10 x	10 x	10 x	2 x	2 x

1

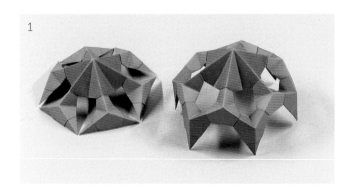

2

3

4

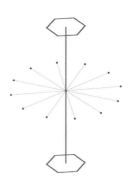

18-Side

First Steps

You need 24 pink triangles (12 x horse and 12 x rider) and 24 turquoise triangles (12 x horse and 12 x rider), as well as 4 moss-green hexagons (2 x horse and 2 x rider). Execute mountain and valley folds according to the illustrations on the right. Join horse and rider to form 12 pink, 12 turquoise, and 2 moss-green modules.

Assembling the Model

Add one turquoise module to each of the six tips of each of the two moss-green modules (fig. 1, *left*). Continue with a 6 x 1 pink module (fig. 1, *right*). This results in 2 halves (fig. 2), which are joined together to form the completed model (fig. 3).

Similarities

S06 (fig. 4, *central*) can be built in several variations: instead of the moss-green hexagonal module, you can also use 6 light-blue triangular modules. Fig. 4 shows variations as a 19-Side (*left*) and as a 20-Side (*right*), here without instructions.

Difficulty: Easy to Moderate

Simple to moderately difficult cutting, simple to moderately difficult construction; time required: approximately 1½ hours.

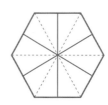
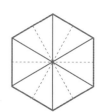

12 x	12 x	12 x	12 x	2 x	2 x

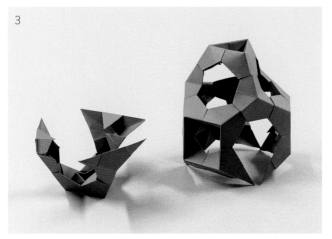

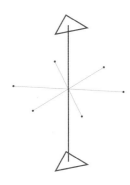

14-Side

First Steps

You will need 36 pink triangles (18 x horse and 18 x rider) as well as 12 turquoise triangles (6 x horse and 6 x rider). Execute mountain and valley folds according to the illustrations on the right. Join horse and rider to form 18 pink modules and 6 turquoise modules.

Assembling the Model

Join 2 x 3 pink modules to form 2 triangular ring joints (fig. 1, *left*). Attach a turquoise module to each open joint (fig. 1, *right*). Join the remaining pink modules to form 3 square ring joints and attach them to the sides (fig. 2, *right*). Join the two parts (fig. 3) to create the finished model (fig. 4).

Difficulty: Easy

Simple cutting, simple construction; time required: approximately 30 minutes.

18 x	18 x	6 x	6 x

1

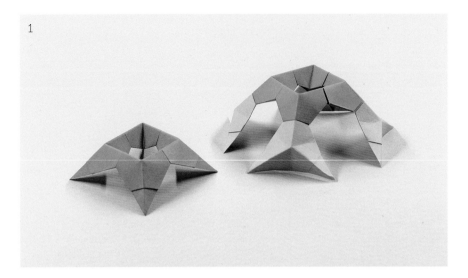

2

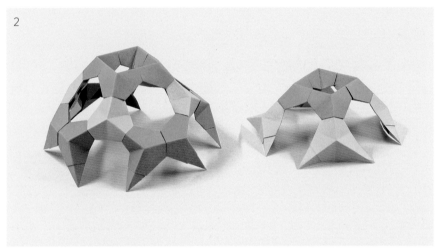

3

4

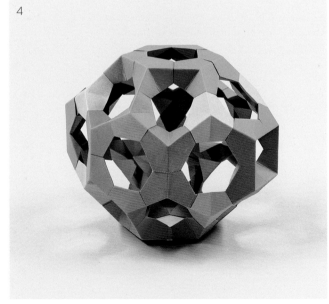

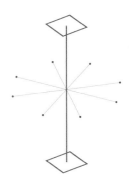

22-Side

First Steps
You need 48 pink triangles (24 x horse and 24 x rider) as well as 16 light-blue triangles (8 x horse and 8 x rider) and 16 turquoise triangles (8 x horse and 8 x rider). Execute mountain and valley folds according to the illustrations on the right. Join horse and rider to form 24 pink, 8 light-blue, and 8 turquoise modules.

Assembling the Model
Join 2 x 4 pink modules to create 2 square ring joints (fig. 1, *left*). Add a light-blue module at each open joint (fig. 1, *right*). These are followed by pink modules as double modules (fig. 2, *left*). The turquoise modules are joined to form double modules and added at the sides of the model (fig. 3, *right*). They are used to put the two parts of the model together to complete the structure (fig. 4).

Difficulty: Easy to Moderate
Simple cutting, simple to moderately difficult construction; time required: approximately one hour.

Alternative Construction
The light-blue triangles can be substituted with pink ones (no illustration).

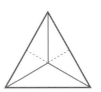

24 x	24 x	8 x	8 x	8 x	8 x

1

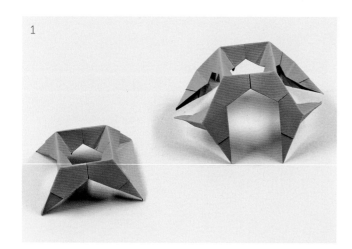

2

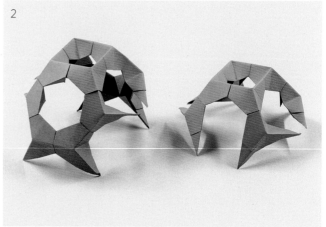

3

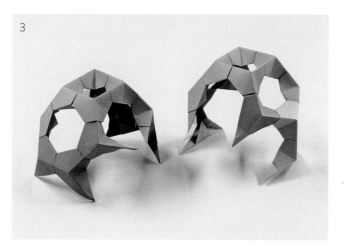

4

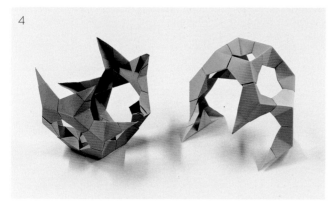

5

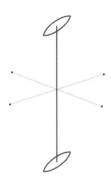

14-Side

First Steps

You need 32 turquoise triangles (16 x horse and 16 x rider) as well as 8 light-blue triangles (4 x horse and 4 x rider) and 8 pink triangles (4 x horse and 4 x rider). Execute mountain and valley folds according to the illustrations on the right. Join horse and rider to form 16 turquoise, 4 light-blue, and 4 pink modules.

Assembling the Model

Join 2 x 4 turquoise modules to form 2 square ring joints (fig. 1, *left*). A further turquoise module is added to each open joint (fig. 1, *right*). Join the light-blue and pink modules to form double modules and attach them to the initial modules at the sides as illustrated in figs. 2 and 3. They form the link between the two other parts of the model (figs. 4 and 5).

Difficulty: Easy to Medium

Simple cutting, simple to moderately difficult construction; time required: approximately one hour.

16 x   16 x   4 x   4 x   4 x   4 x

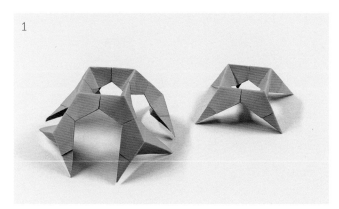

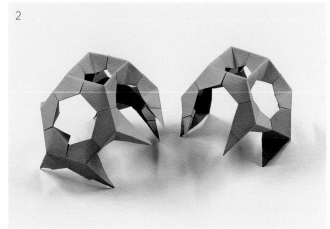

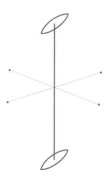

14-Side

First Steps

You need 32 turquoise triangles (16 x horse and 16 x rider) as well as 8 pink triangles (4 x horse and 4 x rider) and 4 dark-blue squares (2 x horse and 2 x rider). Execute mountain and valley folds according to the illustrations on the right. Join horse and rider to form 16 turquoise, 4 pink, and 2 dark-blue modules.

Assembling the Model

Join 2 x 4 turquoise modules to form 2 square ring joints (fig. 1, *right*). Add a further turquoise module to each open joint (fig. 1, *left*). Join the pink modules to form double modules. They are attached to the sides as shown in fig. 2. These, like the two single dark-blue modules, form the link between the two parts of the structure (figs. 2 and 3). Fig. 4 shows the completed model. Fig. 5 shows S10 and S09 next to each other.

Difficulty: Easy to Moderate

Simple cutting, simple to moderately difficult construction; time required: approximately one hour.

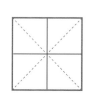
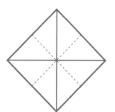

16 x	16 x	4 x	4 x	2 x	2 x

T Models:
From Triangles and Squares
(Mixed)

The models of the T series continue the construction techniques and symmetries of those of the S series. The added use of squares contributes to more colorful models. T01 to T03 form their own series with increasing symmetries (3-fold to 5-fold).

In models T04 through T06, the case is similar (here: 4-fold to 6-fold symmetries). In these models, parts of the construction of the model are variable, as further models show. T07 is a singular form that remains open at two sides (truncated/torso). T08 illustrates the rare case in which two vertices (dark blue in this case) lie crosswise to the main symmetry along the viewed axis. This model has only one 2-fold main symmetry axis and two further 2-fold subsidiary symmetry axes.

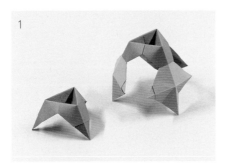

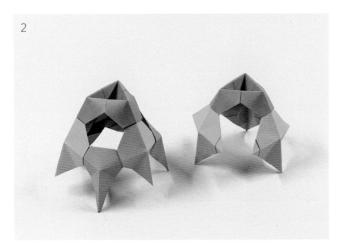

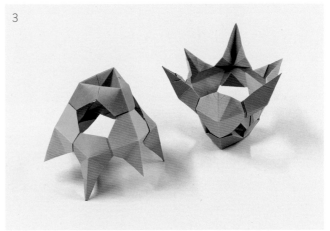

14-Side

First Steps
You need 12 pink triangles (6 x horse and 6 x rider), 12 turquoise triangles (6 x horse and 6 x rider), and 12 light-green squares (6 x horse and 6 x rider). Execute mountain and valley folds according to the illustrations on the right. Then join horse and rider to form 6 pink, 6 turquoise, and 6 light-green modules.

Assembling the Model
2 x 3 modules are joined to form 2 triangular ring joints each (fig. 1, *left*). 3 light-green modules are added to all 3 open joints (fig. 1, *right*), followed by 3 turquoise modules (fig. 2, *left*). This results in 2 identical parts (fig. 3), which are joined to form the complete model. Fig. 4 shows the model from different perspectives.

Difficulty: Easy
Simple cutting, simple construction; time required: approximately one hour.

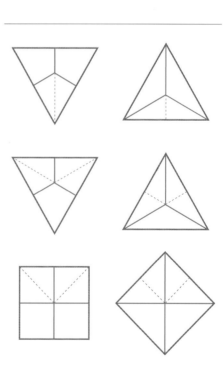

6 x	6 x	6 x	6 x	6 x	6 x

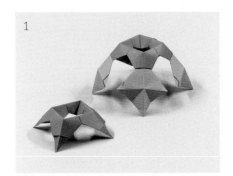

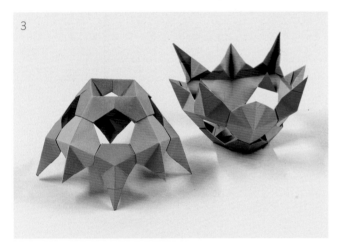

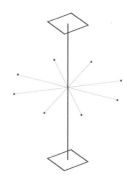

18-Side

First Steps
You need 16 pink triangles (8 x horse and 8 x rider), 16 turquoise triangles (8 x horse and 8 x rider), and 16 light-green squares (8 x horse and 8 x rider). Execute mountain and valley folds according to the illustrations on the right. Join horse and rider to form 8 pink, 8 turquoise, and 8 light-green modules.

Assembling the Model
2 x 4 modules are joined to form 2 square ring joints (fig. 1, *left*). 4 light-green modules are added to each of the 4 free joints (fig. 1, *right*) and then 4 turquoise modules (fig. 2, *left*). This results in 2 identical parts (fig. 3), which are finally put together to form the complete model (fig. 4).

Similarities
The similarities among T01, T02, and T03 are highly obvious (fig. 5).

Difficulty: Easy
Simple cutting, simple construction; time required: approximately 1½ hours.

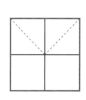 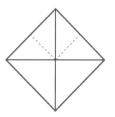

8 x   8 x   8 x   8 x   8 x   8 x

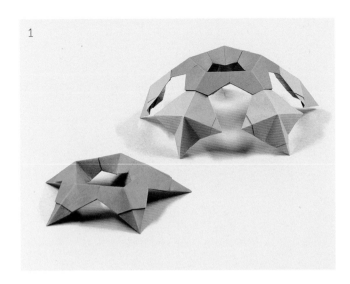

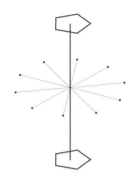

22-Side

First Steps

You need 20 pink triangles (10 x horse and 10 x rider), 20 turquoise triangles (10 x horse and 10 x rider), and 20 light-green squares (10 x horse and 10 x rider). Execute mountain and valley folds according to the illustrations on the right. Join horse and rider to form 10 pink, 10 turquoise, and 10 light-green modules.

Assembling the Model

2 x 5 pink modules are joined to form 2 pentagonal ring joints (fig. 1, *left*). All 5 free joints in each part of the model are continued with 5 light-green modules (fig. 1, *right*) and then with 5 turquoise modules each. This results in 2 identical halves of the model (fig. 2), which are eventually joined (fig. 3).

Similarities

T03 can be continued by using 10 dark-blue square modules to create a 32-Side (here without instructions; see fig. 4, *right*).

Difficulty: Easy

Simple cutting, simple construction; time required: approximately two hours.

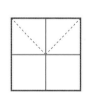
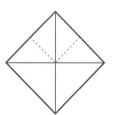

10 x   10 x   10 x   10 x   10 x   10 x

1

2

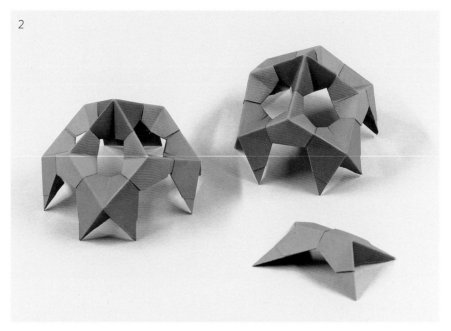

3

4

5

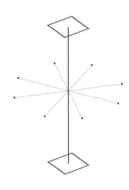

20-Side

First Steps
You need 16 pink triangles (8 x horse and 8 x rider) as well as 16 turquoise triangles (8 x horse and 8 x rider). Additionally, you need 16 orange squares (8 x horse and 8 x rider) and 4 dark-blue squares (2 x horse and 2 x rider). Execute mountain and valley folds according to the illustrations on the right. Join horse and rider to form 8 pink, 8 turquoise, 8 orange, and 2 dark-blue modules.

Assembling the Model
Attach a turquoise module at each of the four ends of each of the 2 dark-blue modules (fig. 1), which are continued in turn with orange modules (fig. 2). This results in 2 identical parts of the model. The pink modules are joined to form double modules (fig. 3), which in turn join the two parts of the model together (fig. 4, *left*; fig. 5, *right*).

Similarities
The 17-Side (fig. 4, *right*; fig. 5, *right*; here without instructions) shows a variant of T04 with 2 different ends. The 2-fold subsidiary symmetries are lost in this variant.

Difficulty: Easy
Simple cutting, simple construction; time required: approximately 1½ hours.

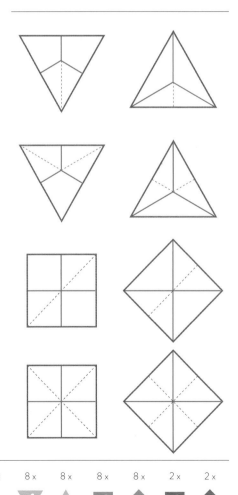

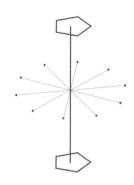

25-Side

First Steps
You will need 20 pink triangles (10 x horse and 10 x rider) as well as 20 turquoise triangles (10 x horse and 10 x rider), an additional 20 orange squares (10 x horse and 10 x rider), and 4 violet pentagons (2 x horse and 2 x rider). Execute mountain and valley folds according to the illustrations on the right. Join horse and rider to form 10 pink, 10 turquoise, 10 orange, and 2 violet modules.

Assembling the Model
Join a turquoise module to each of the 5 points of the 2 violet modules (fig. 1, *left*). Continue with orange modules (fig. 1, *right*). The pink modules as double modules join the two halves (fig. 2) to form the completed model (fig. 3, *left*; fig. 4, *left*).

Similarities
A variation of the 25-Side is the 21-Side (fig. 3, *right*; fig. 4, *right*; fig. 5, *right*). A variation of the 20-Side (T04) is shown in fig. 5 (*left*). Neither variation is explained further here.

Difficulty: Easy to Moderate
Cutting: some challenges, construction: simple; time required: approximately 2½ hours.

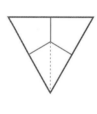

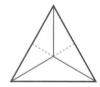
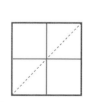
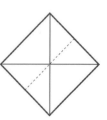
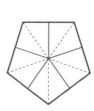
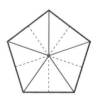

10 x	10 x	10 x	10 x	10 x	10 x	2 x	2 x

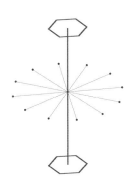

30-Side

First Steps

You need 24 pink triangles (12 x horse and 12 x rider) as well as 24 turquoise triangles (12 x horse and 12 x rider), 24 orange squares (12 x horse and 12 x rider), and 4 moss-green hexagons (2 x horse and 2 x rider). Execute mountain and valley folds according to the illustrations on the right. Join horse and rider to form 12 pink, 12 turquoise, 12 orange, and 2 moss-green modules.

Assembling the Model

Add a turquoise module to each of the 6 points of each of the 2 moss-green modules (fig. 1, *left*), then continue with 6 orange modules (fig. 1, *right*). The pink modules as double modules join the two parts (fig. 2) to form the completed model (fig. 3).

Similarities

The models of entire series (T04, T05, and T06) can be found in figs. 4 and 5, next to each other from different perspectives.

Difficulty: Moderate

Cutting: first challenges, tricky construction; time required: approximately 2½ hours.

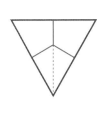

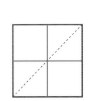
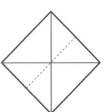

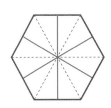
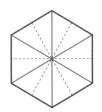

12 x	12 x	12 x	12 x	12 x	12 x	2 x	2 x

1

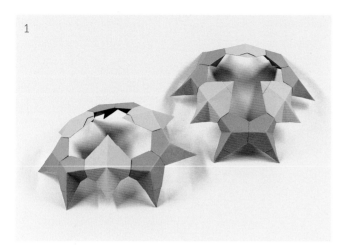

2

3

4

5

24-Side (Truncated)

First Steps

You need 24 pink triangles (12 x horse and 12 x rider) as well as 12 turquoise triangles (6 x horse and 6 x rider). Additionally, 12 orange squares (6 x horse and 6 x rider) and 12 light-green squares (6 x horse and 6 x rider) are needed. Execute mountain and valley folds according to the illustrations on the right. Join horse and rider to form 12 pink, 6 turquoise, 6 orange, and 6 light-green modules.

Assembling the Model

Join pink and light-green modules into a circular sequence to form 2 halves of the model (fig. 1). The pink modules as double modules connect the light-green modules with each other. Attach turquoise modules to the open tips of the light-green modules (fig. 2). Orange modules are added to one of the halves of the model (figs. 3 and 4). Both parts are put together to form a model that is open at two ends (fig. 5). 3 light-green tips remain open at each side.

Difficulty: Moderate

Simple cutting, tricky construction; time required: approximately two hours.

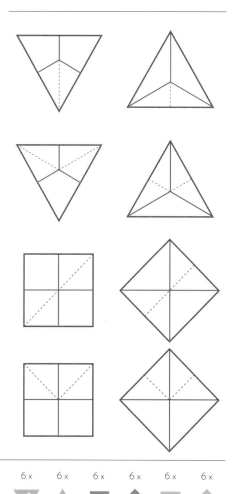

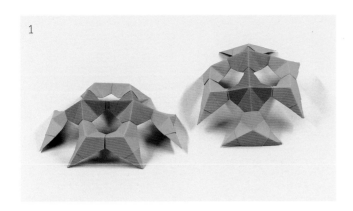

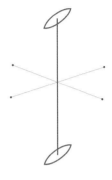

16-Side

First Steps
You need 8 dark-blue squares (4 x horse and 4 x rider) as well as 24 turquoise triangles (12 x horse and 12 x rider) and 8 orange squares (4 x horse and 4 x rider). Execute mountain and valley folds according to the illustrations on the right. Join horse and rider to form 4 dark-blue, 12 turquoise, and 4 orange modules.

Assembling the Model
Join 2 x 2 dark-blue modules to form double modules. A turquoise module follows at each open joint (fig. 1). This results in 2 identical parts of the model, which are attached to a part with 4 orange modules (fig. 2). Fig. 3 shows the completed model.

Difficulty: Easy
Simple cutting, simple construction; time required: approximately 2½ hours.

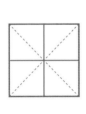 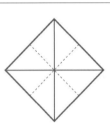

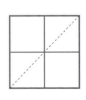 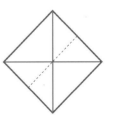

4 x	4 x	12 x	12 x	4 x	4 x

U Models:
From Polygons
(Mixed)

The combinatory possibilities shown in *Folded Forms* culminate in the following models of the U series, where pentagons and hexagons dominate. With this, however, the possibilities are by no means exhausted— this remains for another time and another place*.

U01 and U02 show models with 4- and 6-fold main symmetry axes. Between these could be placed the rhomboid 30-Side**, which, as a high-symmetry form, has six 5-fold symmetry axes.

U03 and U04 have the same construction pathway based on pentagons, differing only in their number and the type of their lateral completion: once in light-blue and once in turquoise modules. U05 and U06 also share the same construction pathway based on dark-blue square modules. Here, pentagons and hexagons vary.

U07 goes back to the modules of the first models but uses them in a different way. U08 and U09 show different solutions with hexagons.

*At the end of this book, connections with *Folding Polyhedra* emerge, thus completing another, bigger circle: among the U models the connections with models from *Folding Polyhedra* are particularly numerous.
**C10 in *Folding Polyhedra*.

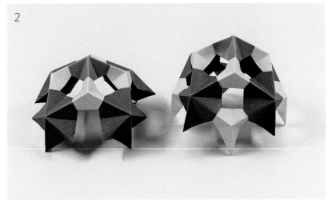

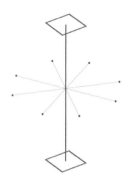

24-Side

First Steps

You need 16 violet (8 x horse and 8 x rider) and 32 light-blue triangles (16 x horse and 16 x rider), as well as 4 dark-blue squares (2 x horse and 2 x rider). Execute mountain and valley folds according to the illustrations on the right. Join horse and rider to form 8 violet, 16 light-blue, and 2 dark-blue modules.

Assembling the Model

Join 2 dark-blue modules with a light-blue module each at each end (fig. 1), then continue with violet modules (fig. 2, *left*). Further light-blue modules continue between the violet modules (fig. 2, *right*). This results in two parts of the model (fig. 3), which are joined in the last step to form the completed model (fig. 4).

Similarities

Fig. 5 shows U01 together with the rhomboid 30-Side*. The two models share obvious similarities.

Difficulty: Moderate

Cutting: some challenges, tricky construction; time required: approximately 1¾ hour.

*C10 in *Folding Polyhedra*.

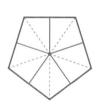 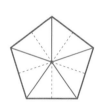

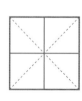 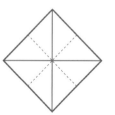

8 x	8 x	16 x	16 x	2 x	2 x

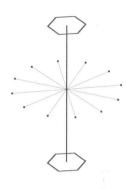

36-Side

First Steps

You need 24 violet pentagons (12 x horse and 12 x rider), 48 light-blue triangles (24 x horse and 24 x rider), and 4 moss-green hexagons (2 x horse and 2 x rider). Execute mountain and valley folds according to the illustrations on the right. Join horse and rider to form 12 violet, 24 light-blue, and 2 moss-green modules. Fix the folded-over tips of the riders underneath the horse with a drop of glue (fig. 1) and keep them in place temporarily with a paper clip.

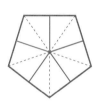
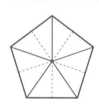

Assembling the Model

Join 2 moss-green modules with a light-blue module each at each vertex (fig. 2, *left*); this is followed by violet modules (fig. 2, *right*). Further light-blue modules follow between the violet modules (fig. 3). This results in 2 identical parts of the model (fig. 4), which are joined in the last step to create the finished model (fig. 5).

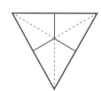

Similarities

Fig. 6 shows next to each other U01 (*left*), U02 (*right*), and the rhomboid 30-Side*.

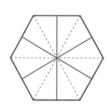
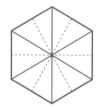

Difficulty: Moderate

Cutting: some challenges, tricky construction; time required: approximately two hours.

*C10 in *Folding Polyhedra*.

12 x	12 x	24 x	24 x	2 x	2 x

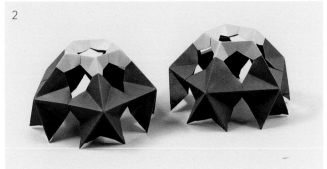

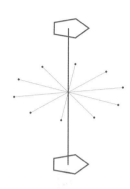

22-Side°

First Steps
You need 20 light-blue triangles (10 x horse and 10 x rider) as well as 20 violet pentagons (10 x horse and 10 x rider). Execute mountain and valley folds according to the illustrations on the right. Join horse and rider to form 10 light-blue and 10 violet modules.

Assembling the Model
Join 2 x 5 light-blue modules to form 2 pentagonal ring joints (fig. 1, *left*). This is followed by violet modules (fig. 1, *right*). This results in 2 identical halves of the model (figs. 2 and 3), which are joined in the last step to form a finished model (fig. 4).

Similarities
Fig. 5 shows on the lefthand side a highly symmetrical icosahedron and, on the righthand side, a highly symmetrical dodecahedron*. U03 is a mix of the two.

Difficulty: Moderate
Cutting: first challenges, tricky construction; time required: approximately 1¾ hour.

*B08 and B04 in *Folding Polyhedra*.

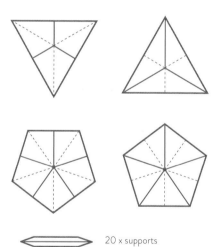

20 x supports

Note:
The supports stabilize the joints between two neighboring pentagonal modules. The pentagonal horse and rider papers have deviating measurements (see https://www.haupt.ch/faltformen).
°Some folding measurements deviate from the standard measurements (see page 18).

10 x	10 x	10 x	10 x

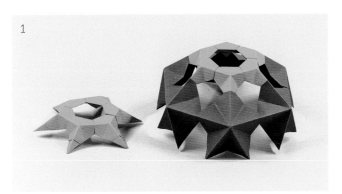

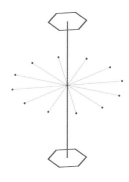

26-Side°

First Steps

You need 24 turquoise triangles (12 x horse and 12 x rider) as well as 24 violet pentagons (12 x horse and 12 x rider). Execute mountain and valley folds according to the illustrations on the right. Join horse and rider to form 12 turquoise and 12 violet modules.

Assembling the Model

Join 2 x 6 modules to form ring joints (fig. 1, *left*). Add violet modules to these (fig. 1, *right*). This results in 2 identical parts of the model (figs. 2 and 3), which are joined in the last step (fig. 4).

Similarities

Figs. 5 and 6 show U03 and U04 from different perspectives.

Difficulty: Easy to moderate

Cutting: some challenges, simple construction; time required: approximately one hour.

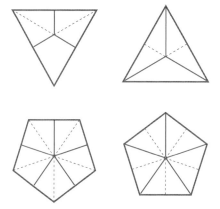

24 x supports

Note:

The supports stabilize the joints between two neighboring pentagonal modules. The pentagonal horse and rider papers have deviating measurements (see https://www.haupt.ch/faltformen).
°Some folding measurements deviate from the standard measurements (see page 18).

12 x	12 x	12 x	12 x

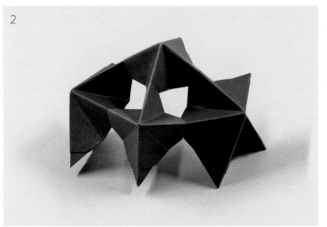

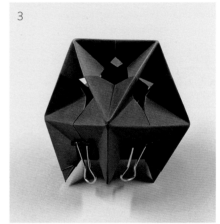

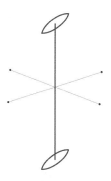

12-Side° (Oloidal)

First Steps

You need 8 dark-blue squares, consisting of 4 x horse and 4 x rider, as well as 8 violet pentagons with 4 x horse and 4 x rider. Execute mountain and valley folds according to the illustrations on the right, and join to form 4 dark-blue and 4 violet modules. The folded-over tips of the riders are glued underneath the horse (fig. 1) and temporarily fixed with clips.

Assembling the Model

2 dark-blue modules are put together. These are continued with violet modules on all sides (figs. 2 and 3). Some glue helps fix the joint permanently in place. The remaining 2 dark-blue modules form the capstones for the model (figs. 4 and 5).

Difficulty: Moderate

Cutting: some challenges, tricky construction; time required: approximately one hour.

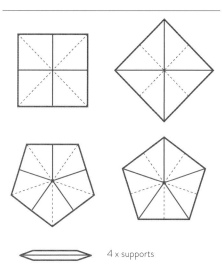

4 x supports

Note:

The supports stabilize the joints between two neighboring pentagonal modules. The pentagonal horse and rider papers have deviating measurements (see https://www.haupt.ch/faltformen).

°Some folding measurements deviate from the standard measurements (see page 18).

4 x   4 x   4 x   4 x

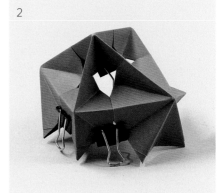

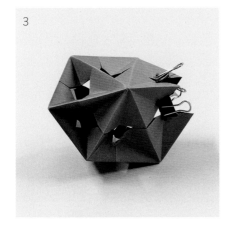

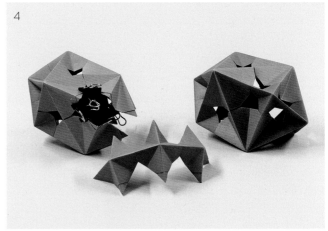

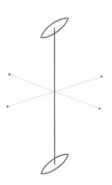

16-Side (Oloidal)

First Steps
You need 12 dark-blue squares (6 x horse and 6 x rider) and 8
moss-green hexagons (4 x horse and 4 x rider). Execute mountain
and valley folds according to the illustrations on the right. Join
horse and rider to form 6 dark-blue and 4 moss-green modules.
The folded-over tips of the riders are glued to the underside of
the horse (fig. 1) and temporarily fixed with clips.

Assembling the Model
Join 3 dark-blue modules to form a row and continue with
moss-green modules all around (figs. 2 and 3). Glue helps fix
the joints permanently in place. The remaining 3 dark-blue
modules form the capstones for the model (fig. 4).

Similarities
Fig. 5 shows the octahedron* as well as U05 and U06 alongside
each other.

Difficulty: Moderate
Cutting: some challenges, tricky construction; time required:
approximately one hour.

*A01 in *Folding Polyhedra*.

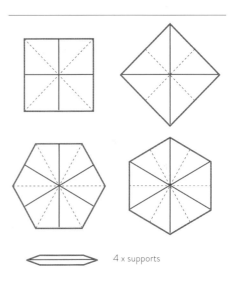

4 x supports

Note:
The supports stabilize the joints between
two neighboring hexagonal modules.

6 x   6 x   4 x   4 x

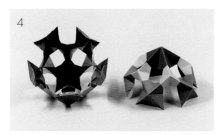

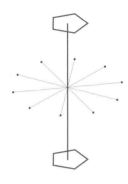

20-Side

First Steps
You need 4 violet pentagons (2 x horse and 2 x rider), 20 light-blue triangles (10 x horse and 10 x rider), and 20 dark-blue squares (10 x horse and 10 x rider). Execute mountain and valley folds according to the illustrations on the right. Join horse and rider to form 2 violet, 10 light-blue, and 10 dark-blue modules. Glue the folded-over tips of the riders of the violet and the dark-blue modules in place underneath the horse (fig. 1) and temporarily hold in place with clips.

Assembling the Model
Join the 2 violet modules at each vertex with a light-blue module (fig. 2, *left*). These are followed all the way around with the dark- blue modules (fig. 2, *right*). This results in two identical halves of the model (figs. 3 and 4), which are joined to complete the model in the last step (figs. 5 and 6).

Difficulty: Easy to Moderate
Cutting: some challenges, easy construction; time required: approximately one hour.

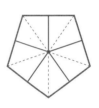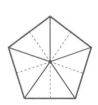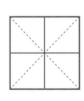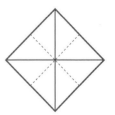

2 x   2 x   10 x   10 x   10 x   10 x

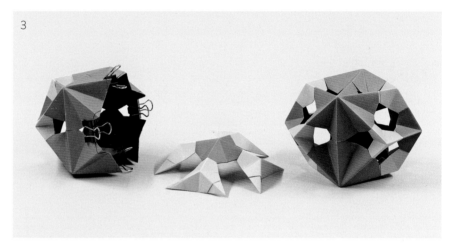

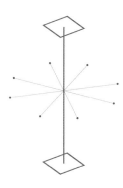

16-Side

First Steps

You need 4 dark-blue squares (2 x horse and 2 x rider), 16 light-blue triangles (8 x horse and 8 x rider), and 8 moss-green hexagons (4 x horse and 4 x rider). Execute mountain and valley folds according to the illustrations on the right. Join horse and rider to form 2 dark-blue, 8 light-blue, and 4 moss-green modules. Glue the folded-over tips of the riders of the dark-blue and moss-green modules in place underneath the horse (fig. 1) and fix temporarily with clips.

Assembling the Model

Join the 2 dark-blue modules at each vertex with a light-blue module each. This results in 2 identical parts of the model (fig. 2). Add the moss-green modules to one of these and glue the joints together permanently (fig. 3, *left*). Take the other part of the model (fig. 3, *center*) and join it to the first in a last step. Fig. 4 shows the finished model.

Difficulty: Moderate

Cutting: some challenges, tricky construction; time required: approximately one hour.

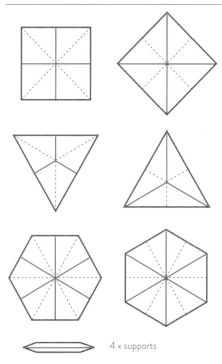

4 x supports

Note:

The supports stabilize the joints between two neighboring hexagonal modules.

2 x	2 x	8 x	8 x	4 x	4 x

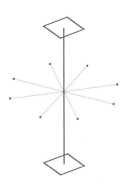

18-Side

First Steps

You need 16 dark-red squares (8 x horse and 8 x rider) as well as 8 moss-green hexagons (4 x horse and 4 x rider). Execute mountain and valley folds according to the illustrations on the right. Join horse and rider to form 8 dark-red and 4 moss-green modules. Glue the folded-over tips of the riders of the dark-red and moss-green modules in place underneath the horse (fig. 1) and fix them temporarily in place with clips.

Assembling the Model

Join 2 x 4 dark-red modules to form a square ring joint each (fig. 2, *right*), and glue these together (fig. 2, *left*). This results in two identical parts of the model. Join the moss-green modules to one of these and glue them in place (fig. 3, *center*). Add the second dark-blue part to create the finished model (fig. 3, *right*).

Similarities

Fig. 4 shows the similarities between U08 and U09.

Difficulty: Easy to Moderate

Cutting: some challenges, simple construction; time required: approximately one hour.

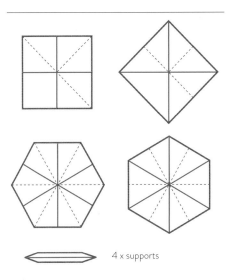

4 x supports

Note:
The supports stabilize the joints between two neighboring hexagonal modules.

8 x   8 x .   4 x   4 x

Appendix

Didactic Hints

The forms included in *Folding Polyhedra* and *Folded Forms* offer a large selection of folded spatial solids. But this does not mean that all the possible forms that can be conceived and folded have been exhausted yet!

In this book, we can't forget the authors Rona Gurkewitz and Bennett Arnstein, who pioneered (among other things) the horse-and-rider technique—without naming it as such. They were probably the first to create an entire range of models in this technique. The approach and a number of technical solutions in both books are, however, my own and go far beyond the explanations by Gurkewitz and Arnstein.

General Considerations
Paper folding speaks to people in several different ways. Depending on the impulse and context, it can be used as a focal point at home, in teaching, at university, or for therapeutic reasons.

Meditative Quality of Working on a Repetitive Task
The continuously repeating and repetitive folding of polygons and the joining of modules is experienced by many people as a type of practical meditation. This is particularly the case with those models that consist of only one simple type of module. In this book, there are only a few models of this kind (Q models). In *Folding Polyhedra*, the models from chapters A and B are particularly

well suited to this kind of approach. Here you can experience in a special way that the finished model is qualitatively "more" than just a sum of its parts, and this experience is the reward for the work you have put in. And it is not at all necessary that you understand which geometrical conditions have to be fulfilled so that a particular folded form can be created.

From Experimenting and Combining to Understanding
Once the interest in folding polyhedra has been kindled in the way indicated above, the next steps suggest themselves naturally—to try out which other forms can be realized in this way of working. To experiment and combine successfully, you need a certain free space, in terms of having both enough time as well as inward space. You also need enough patience when you meet with initial problems: as a teacher, you may want to withhold certain solutions in order to stimulate experimentation, or you may like to give specific tasks (e.g., limiting the experiments to certain folds or modules or quite specifically to certain models).

Gradually and almost without noticing, you will gain certain insights; for example, why certain combinations fit, or indeed why they don't. To understand this with exact certainty requires strict geometric-mathematical insights. On the other hand, paper is an elastic material that allows for the formation of spatial forms, which according to strict geometrical rules cannot be constructed

Joining the Modules

from planes. In this book especially, there are many such forms. In some, this is obvious through the visibly curved ring joints (faces); in others, this is not immediately obvious.

Craft Aspects

Folding polyhedra requires certain fine motor skills and craft abilities. It can in this way become a serious point to examine one's own abilities. In the summer of 2020, an entire cohort of applicants for a degree in dentistry at the University of Witten/Herdecke (Germany) were subjected to a test of the care they gave to a craft task. Only slightly more than half the applicants passed the test. The same test was given to students further on in their course. It emerged that the students' success in folding polyhedra correlated largely with other observations on their dexterity and fine motor skills. Since dentistry requires a high degree of fine motor skills, it is obvious that folding polyhedra is a meaningful test to identify suitable applicants. This can be applied to a number of professions. Initially, folding polyhedra does not require any extraordinary intellectual or cognitive abilities as reflected in school grades. It can be learned by going through the process itself ("learning by doing"). You may want to test your dentist prior to your next appointment to find out whether he can fold a polyhedron properly!

Experiencing Space

One of the appeals of spatial experience lies in the fact that a spatial body (a solid) can be looked at and touched from different sides. With folded polyhedra, the journey begins with level (plane) polygonal pieces of paper. Once you start folding, they rise from the plane into space. This touches on a significant problem in grasping spatial forms, and that is the development from the second dimension (plane) to the third dimension (space). When you fold polyhedra, this is a step that happens by itself and thus teaches about the experience of this change.

This experience of space relies especially on immediate haptic experiences, although seeing (as visual perception) is directly involved too. A particular quality of spatial experience can be grasped when you fold polyhedra with your eyes closed.

By actively experiencing something with our own hands, we can feel something of the joy and the "dignity of the unknown," as Novalis put it in his contemplations on romanticism. As humans, we need to experience the unfinished, the incomplete, to feel we have not yet understood or completed everything, but to continue to develop something and to be able to creatively continue Creation. In geometry, a large number of spatial forms have been known for a long time—but it is still possible to find new aspects in them. This, too, is something that this book tries to inspire.

The joints can be further stabilized with a drop of glue.

Done: the assembled works in the school window

In School

In the context of many school subjects, folded forms can be an exciting and fulfilling project: mathematics, art, design, and technology, as an activity for the last lesson before vacation, or for substitute teachers. My experience in teaching design and technology lessons (focus: carton and bookbinding) has shown that folded forms can be a rewarding project for many pupils.

It often makes sense to start with models using only a single type of modules; for example, with the models from the Q series*. Students with a lot of initiative may have to be steadied somewhat so that they don't race ahead. It is also important to devote particular care to explaining every step in great detail, breaking each step down into a number of smaller steps. Only when the first steps have been checked and, if necessary, corrected should the students start on the next step. Otherwise, it becomes difficult to keep up with corrections. If you have some students who grasp things quickly, you can ask them to help others. Once an introductory period has passed, it might also make sense to give such students more difficult special tasks.

When the first models have been completed successfully, some students will have had enough for the moment and will come back to folded forms only later in life. On the other hand, there are always some students—even among the younger ones—who take on folded forms as their own project and who enjoy working on them on their own over a longer period of time and even at home. For such students it is often quite enough to keep sufficient folding paper on hand and to be able to instruct them on the spot as to which folded forms will yield a certain result. This book offers much visual material to assist the teacher. Most models are easy to realize**. In preparation, it has proven helpful to make the visual representations of horse, rider, mountain, and valley folds really one's own and to be able to teach these in a lesson quickly and securely. The same can be said of the information on the symmetry axes, which can be integrated into a practical lesson on "design with paper" in its science aspect— in the sense of an interdisciplinary approach to teaching.

In combining different basic shapes of paper, a more playful element comes in: using colored paper in combination with different forms of folding is emotionally appealing and motivates many people to try out which modules can be combined in such a way that a closed form results. And if things do not fit, it is still rewarding to find out why this is the case. The result in each case always immediately arises from experimentation. Additionally, an intellectual-cognitive process of understanding is stimulated by conscious questioning. In this way, the general insight is gained that through practical failure, meaningful understanding is acquired, which may already be usefully applied the next practical experiment and may well lead to a successful outcome.

Collected works

Folded lampshade (model: Johanna Bolte, Luisa Rüschenschmied)

A popular question is that concerning the possible size of folded forms: Just how big can you make them? Answering this question in a practical manner also addresses a number of thematic questions. As a basic rule, you can enlarge all measurements nearly infinitely as long as they remain in proportion. With increasing size, the proper weight comes into play in relation to the stability of the paper. The 36-Side (R07) or the 22-Side (S03)*** when built from triangles with an edge length of 7 inches (18 cm) will have a diameter of approximately 16.5 ft. (0.5 m). Even when built with paper weighing 200g /m², all joints of models of this size have to be glued in order to be stable. If you do not do this, horse and rider as well as neighboring modules will slide out of place due to the weight they carry.

Folding the models from this book provides an excellent opportunity to get to know different types of paper and their respective advantages and disadvantages. Stiff, hard paper for transparencies folds easily but often breaks when it is folded too sharply. Cardboard, because of its thickness, provides some resistance when folded (I recommend the use of a folding bone), but it will carry larger forms (and higher weight).

For open days and parents evenings or other public events, folded forms make for a charming, expressive eye-catcher.

Math lessons can easily connect to the folded forms, using them to practice calculating area, volume, and usage: "How many triangles of size x can I cut from a sheet of paper of the size y?," or "How many sheets of paper of the size of y do I need, if I want z number of models with a certain number of horse and rider pieces of size x?"

* Or the models from the A and B series from *Folding Polyhedra* / **In *Folding Polyhedra*, you will find very demanding projects alongside very easy ones / *** Or a truncated icosahedron (B07), or the extended truncated icosahedron (L06) from *Folding Polyhedra*

Glossary

BASIC TERMINOLOGY FOR FOLDED FORMS

Simple and complex symmetries. Folded forms have simple symmetries if they have one main symmetry axis; they may also have several subsidiary symmetry axes (in short: main and subsidiary symmetry axes). Models with two separate and multi-fold symmetry axes and additional 2-fold symmetry axes are far more complex.

Main (symmetry) axis / subsidiary (symmetry) axis. Many models in this book have a main symmetry axis with a high number of symmetries, which is intersected laterally by several 2-fold axes. For practical reasons, these axes are called main and subsidiary axes.

Model, module, double module, free tips. Each model consists of a number of modules. Each module consists of two pieces of paper in the shape of a regular polygon. The upper piece is here called rider, the lower horse.

Some tips of the rider will protrude from the module. These form the joints with neighboring modules. As long as the joint has not been put together, it is open (see also page 10, *figure left and center*).

Polyhedron. (Greek: *poly*: many; *hedron*: area, face). In general: a solid with many faces. Example: cube. Many-faced solids are named after the number of their ring joints in *Folded Forms*. Example: an 18-Side designates a folded form with 18 ring joints.

Polygon. (Greek: *poly*: many; *gon*: vertex, edge). In general: figure with multiple edges. Example: square.

Ring joints. Modules formed of horse and rider as shown and explained in the different instructions. Several modules (a minimum of 3) form an angular ring joint (on this, see also page 10, *figures center and right*). In this book, you find models with 3-, 4-, 5-, and 6-fold ring joints. The sum of the ring joints determines the name of the model. Example: If a model has 18 ring joints, it is in this book called an 18-Side.

Tubular. Several ring joints together can form a part (torso) of a model that consists of a tubular combination of several ring joints. For the benefit of clear terminology, the term "tubular" is used here, not "second-grade ring joint" or ring joint of "second/higher order."

Symmetry axes. An imagined straight line through the center of the solid, which at the same time forms an axis for a spatial rotation. In symmetrical spatial forms (here: folded forms), a symmetry axis runs not only through the center of the model, but also through two opposing vertices, through the center of opposing faces, or through the center of the edges (for more on this, see page 11).

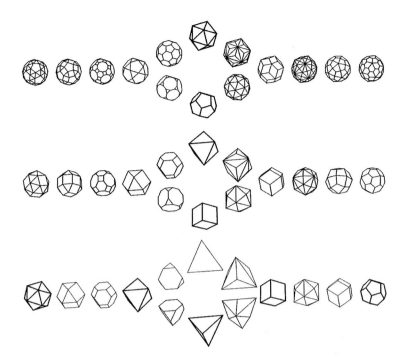

All polyhedra forms at a glance:
in the central line: Platonic solids, left side:
Archimedean solids, right side: Catalan solids (figure
by Helmut Emde, rearranged; see bibliography, page
169).

Topology. (Greek: *tópos*: form/*gestalt*) The teaching of the appearance and characteristics of forms.

Torso. (truncated solid; from Latin: *thyrsus*; Greek: *thursos*, stump). Originally a term used for the (predominantly) artistic rendering of a body in which major limbs are missing. Here it is used for models that for geometric or practical reasons are not realized, with all ring joints completed and that are therefore open at two ends.

n-foldness. During a 360° rotation around a symmetry axis of a spatial solid, the same view appears twice. This is an example of a 2-fold symmetry axis. The models in this book can also be 3-, 4-, 5-, 6-, 7-, or 8-fold. Ring joints, too, always have a certain numbered foldness.

TERMS DESCRIBING FORM

Chiral. Mirror symmetrical. Chiral spatial forms can be realized in two mirror symmetrical variations that are related to each other, such as the right hand to the left hand. Example: 18-Side (model O09 on page 50).

Regular. Regularly formed. A polygon is regular if all sides have the same length and if all angles between the edges are equal. A polyhedron is regular if all edges are of the same length, all faces are ring shaped, and the angles between neighboring faces are equal. All Platonic solids are regular.

Semiregular. Only partially regularly formed. A polyhedron is semiregular if the faces are equal, but the angles at the vertices are of different sizes (Catalan solids). Alternatively, it can consist of different faces, but the edges are of the same length (Archimedean solids).

Platonic solid. Synonym for regular polyhedra. The five regular spatial forms named after Plato (ca. 428–347 BCE) are tetrahedron, cube, octahedron, dodecahedron, and icosahedron.

Archimedean solid. A total of 13 semiregular polyhedra (plus two chiral forms) were named after the Greek mathematician Archimedes (ca. 287–212 BCE). Each Archimedean solid can be traced back to one or two Platonic solids. Each Archimedean solid also has a Catalan dual partner.

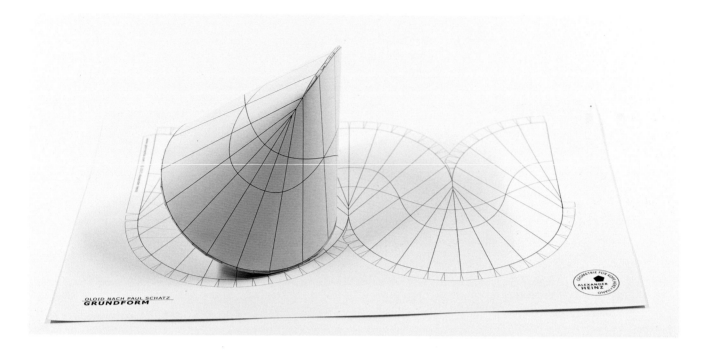

Catalan solid. A total of 13 semiregular polyhedra (plus two chiral forms), which were named after the Belgian mathematician Eugène Catalan (1814–1894). Each Catalan solid can be traced back to one or two Platonic solids, which also explain the names of the Catalan solids. Each Catalan solid has an Archimedean dual partner.

Oloid. The oloid is a form with two curved edges that are enclosed by only one curved face. Its two curved edges face each other crosswise. Oloidal folded forms have two edges each, which also face each other crosswise (see figure above).

Oloidal. Topological similarities with the form of the oloid.

Bibliography and Image Credits

BIBLIOGRAPHY

Adam, Paul, and Arnold Wyss. 1994. *Platonische und Archimedische Körper, ihre Sternformen und polaren Gebilde.* 2nd ed. Stuttgart: Verlag Freies Geistesleben.

Emde, Helmut. "Zur Geometrie räumlicher Strukturen." In *Diatomeen I: Schalen in Natur und Technik*, 222–243. Institut für leichte Flächentragwerke, University of Stuttgart, IL28.

Gurkewitz, Rona, and Bennett Arnstein. 2003. *Multimodular Origami Polyhedra. Archimedeans, Buckyballs and Duality.* Mineola, NY: Dover.

Heinz, Alexander. 2021. *Folding Polyhedra: The Art & Geometry of Paper Folding.* Atglen, PA: Schiffer.

Heinz, Alexander, and Andreas Vahlenkamp. "Polyederfalten als Auswahltest: Studienplatzbewerbung für Zahnmedizin an der Universität Witten/Herdecke." In *Informationsblätter der Geometrie* IBDG 2/2020, vol. 39. Editor: Fachverband der Geometrie (ADG, Österreich).

Heinz, Alexander. "Falt-Polyeder: Eine west-östliche Verbindung." In *Informationsblätter der Geometrie* IBDG 2/2012, vol. 31. Editor: Fachverband der Geometrie (ADG, Österreich).

Offermann, Erich. 2004. *Kristalle und ihre Formen.* Achberg, Germany: KristalloGrafik Verlag.

Reiners, Ludwig. 1974. *Ein Volksbuch der Dichtung.* Munich: C. H. Beck.

IMAGES

Page 14/15:
Adam, Paul, and Arnold Wyss (see Bibliography)

Page 167:
Emde, Helmut (see Bibliography)

Didactic Hints:
Barbara Jordak (colored models)
Johanna Bolte (lampshade)

Alexander Heinz

Alexander Heinz (*1968) is a master bookbinder and studied art (B. Ed.) at TU Dortmund. He teaches bookbinding and geometry at a school in North-Rhine Westphalia (Germany), conducts workshops for young people and adults, and presents lectures at universities, colleges, and other institutions of adult education in Germany, Austria, and Switzerland. He has written numerous articles on polyhedra forms and other subjects of geometry. With his free project approach to building models, he combines craftsmanship, art, and geometry.

Publications (articles):

- "Polyederfalten als Auswahltest: Studienplatzbewerbung für Zahnmedizin an der Universität Witten/Herdecke" ("Polyhedral Folds as a Selection Test: Study Place Application for Dentistry at the University of Witten/Herdecke"). IBDG, 2/2020, information sheets of the Geometry, Communication Organ of the Austrian Fachverband der Geometrie, ADG.
- "Aus der Zeit gefallen: Polyeder als momentane Durchgangsstationen von Transformationen. Ein Beitrag zur Polyeder-Morphogenese" ("Fallen Out of Time: Polyhedra as a Momentary Transit Station of Transformations: A Contribution on Polyhedron Morphogenesis"). IBDG, 1/2020.
- "Nach den Sternen greifen: Räumliche Sternbilder" ("Reaching for the Stars: Spatial Constellations"). IBDG, 1/2018.

- "Ist die Erde ein Stern? Geometrische und geologische Sicht auf die Erdform" ("Is the Earth a Star? Geometric and Geological View of the Earth's Shape"). IBDG, 2/2017.
- "Das Runde muss ins Eckige: Ballformen und ihre Grundlagen" ("The Round Has to Go into the Square: Ball Shapes and Their Basics"). IBDG, 1/2017.
- "Sonne, Monde, Wandelsterne: Ein Planetarium aus astronomischen Bastelbogen" ("Sun, Moons, Walking Stars: A Planetarium of Astronomical Handicraft Sheet"). IBDG, 2/2013.
- "Faltpolyeder: eine west-östliche Verbindung" ("Folding Polyhedron: A West-East Compound"). IBDG, 2/2012.
- "Mit Krummen und Geraden auf der Überholspur" ("With Crooks and Straights in the Fast Lane"). erziehungsKUNST, 3/2011, pp. 32–35.
- "Geometrie in Bewegung: 80 Jahre Schatz'sche Umstülpung" ("Geometry in Motion: 80 Years of Treasure Hunting"). IBDG, 1/2010.
- "Ein Stein kommt ins Rollen. Oloid-Woche in Basel" ("A Stone Starts Rolling: Oloid Week in Basel"). IBDG, 2/2009.
- "Platonische Körper erleben. Gast-Unterricht in einer 8. Klasse" ("Experience Platonic Bodies: Guest Lessons in an 8th Grade"). erziehungsKUNST, 7/8/2007, pp. 812–815.
- "Kulturgeschichtliche und geometrische Aspekte zur Entwicklung des Raumbewusstseins" ("Cultural-Historical and Geometric Aspects of Development of Spatial Awareness") .Human and Architektur, 11/2007, pp. 58–63.

Subject Focus:

Regular and semiregular polyhedra as points in processes of development (morphogenesis of polyhedra).

Retro-perspective, camera obscura / laterna magica, impossibles (optical illusions).

Eversion and models that can be turned inside out, the oloid and other space-time forms.

Polyhedra in daily life and the cultural history of polyhedra.

Models:

- Regular and semiregular polyhedra, ball polyhedra, oloid, and similar forms.
- Movable forms and forms that can be turned inside out
- Walk-in forms
- Hand models (didactic)
- Presentation models (didactic, industrial fairs)
- Craft kits
- Moving images

Courses, Lectures, and Exhibitions:

Recklinghausen Observatory, Roemer-und-Pelizaeus Museum Hildesheim, University of Freiburg i. Br., PH Fribourg, University of Innsbruck, PH Steiermark/Graz, TU Graz, PH Salzburg, PH Kärnten/Klagenfurt, OLMA Fair St. Gallen. In the context of conferences of the DGfGG (German Society for Geometry and Graphic Design): TU Munich, TU Dresden, KIT (previously Karlsruhe University).

More information at:

www.geomenta.com

Acknowledgments

I would like to thank the Haupt Verlag, especially Ms. Heidi Müller, for the positive collaboration. I thank Frank Georgy for reliable and good coordination with regard to the book's design. Thanks to all my readers of *Folding Polyhedra* for their friendly and constructive feedback and suggestions.

For the English edition, I would like to thank Schiffer Publishing for taking on this second volume and the continued positive collaboration. I would also like to thank Katrin Binder for her translation and Neil Franklin for editing the same.

Over many years, I have collaborated with colleagues from the Austrian Society for Geometry (ADG), and I owe them a debt of gratitude for many inspirations. I would like to mention especially the exchanges at their annual convention in Strobl and the Day of Geometry at TU Graz. My particular thanks go to Friedhelm Kürpig for the preface. For several helpful suggestions and support, I thank Günter Maresch (University of Salzburg).

My meetings with lecturers and subject teachers of mathematics were always an enriching experience. I have to mention the University of Karlsruhe (now KIT), TU Dresden and TU Munich, the University of Freiburg/Breisgau, and the PH Fribourg (CH). To many students and lecturers in the Department of Arts at the TU Dortmund, I owe my thanks for many deep and far-reaching conversations on the subject, as well as on the connections among craftsmanship, art, and science.

A special thanks is due to my family: to my parents, who allowed me to grow up in an environment characterized by design and art, and to my wife and children, who have carried my enthusiasm for many years with much understanding and patience through all its ups and downs. Michael Doman (Kampen/Sylt) has accompanied my work with warm interest over many years and contributed much valuable advice and lively exchanges of thoughts on the geometry of polyhedra. Thank you! To Martin Mahle (Eichenau) and Bruno Hoffman (Lehrte), I owe original observations and suggestions.

For helpful meetings and exchanges, I thank Gert Hansen (Copenhagen), Ueli Wittorf (Zurich), Fred Voss (Hannover), Jürgen Blasberg (Hagen), Christoph Bednarz (Bochum), Alexander Junge (Berlin), Niels Junge (Hannover), and Rita Baumgart† (Essen).

Finally, my gratitude also goes to my teachers at the Rudolf-Steiner-Schule Dortmund who taught me the first rich basics of geometry—those teachers who have died and those who are still alive. I equally thank my students at this school who took up my folded polyhedra with lively enthusiasm and of whom some even took the subject further by themselves.

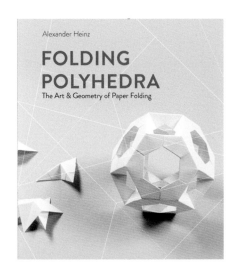 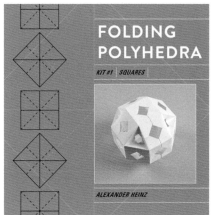 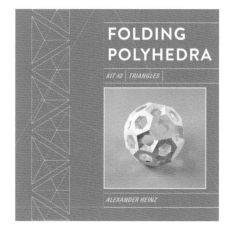

Folding Polyhedra: The Art & Geometry of Paper Folding
Schiffer Publishing, 2021
ISBN: 978-0-7643-6157-9
https://schifferbooks.com

Explore the reaches of your imagination by combining art, craft, and mathematics to create the age-old geometric form of the polyhedra. More than 50 projects use just one center point to create shapes that are folded symmetrically and build upon each other. With easy-to-follow, step-by-step instructions, you can construct models from regular polygons, including equilateral triangles, squares, pentagons, hexagons, octagons, and decagons. These engaging projects require active, mental participation and can aid in reshaping familiar thought patterns as well as keeping your focus on the present moment as a form of relaxation. Also included is a wealth of detail on the subject of geometry. Once a requirement for entry into the philosopher Plato's Academy, now geometry can be enjoyed by everyone with an eye for fun. The book includes a chapter of tips to create your own designs.

Folding Polyhedra Kit 1
Squares
Schiffer Publishing, 2022
ISBN: 978-0-7643-6273-6
https://schifferbooks.com

Folding Polyhedra Kit 2
Triangles
Schiffer Publishing, 2022
ISBN: 978-0-7643-6274-3
https://schifferbooks.com

Folding Polyhedra Kit 3
Triangles and Squares
Schiffer Publishing, 2022
ISBN: 978-0-7643-6312-2
https://schifferbooks.com

Folding Polyhedra Kit 4
Multi-Triangles
Schiffer Publishing, 2022
ISBN: 978-0-7643-6311-5
https://schifferbooks.com

Other Schiffer Books by the Author:
Folding Polyhedra: The Art & Geometry of Paper Folding, ISBN 978-0-7643-6157-9

Other Schiffer Books on Related Subjects:
Paper Sculpture: Fluid Forms, Richard Sweeney, ISBN 978-0-7643-6214-9
Pop-Up Paper Spheres: 23 Beautiful Projects to Make with Paper and Scissors, Seiji Tsukimoto, ISBN 978-0-7643-6429-7

Originally published as *Faltformen: Papierdesign zwischen Symmetrie und freiem Spiel* by Alexander Heinz, copyright © 2021, Haupt Verlag, Bern
Translated from German by Katrin Binder

Library of Congress Control Number: 2022944472

Text and photos, unless otherwise stated:
Alexander Heinz, DE-Herdecke, geomenta.com
Design and typesetting: Frank Georgy (Kopfsprung.de), DE-Cologne, with the assistance of Alexander Heinz
Illustrations: Alexander Heinz

Cover design by Jack Chappell
Type set in Brandon Grotesque

ISBN: 978-0-7643-6612-3
Printed in China

Published by Schiffer Publishing, Ltd.
4880 Lower Valley Road
Atglen, PA 19310
Phone: (610) 593-1777; Fax: (610) 593-2002
Email: Info@schifferbooks.com
Web: www.schifferbooks.com

For our complete selection of fine books on this and related subjects, please visit our website at www.schifferbooks.com. You may also write for a free catalog.

Schiffer Publishing's titles are available at special discounts for bulk purchases for sales promotions or premiums. Special editions, including personalized covers, corporate imprints, and excerpts, can be created in large quantities for special needs. For more information, contact the publisher.

We are always looking for people to write books on new and related subjects. If you have an idea for a book, please contact us at proposals@schifferbooks.com.